Breadmaking for the Beginners

初嘗麵包香

獨角仙 著

為仙姐寫序，真是有點壓力。執筆前腦海不停地思索，應該要寫些甚麼。因為讚美之詞，相信大家已經寫得太多或看得太多了。最後，我決定將這個寫序的機會，看成寫紀念冊一樣，記錄一下我跟仙姐之間的一些事情。

認識仙姐是從看她的 BLOG 開始，發覺大家都非常擁護她；後來有機會跟朋友去參加她的麵包課程，發覺她真的有過人魅力之處。她對烘焙的熱誠，及跟別人無私分享烘焙知識的熱情，令我十分佩服。

在一個偶然的機會下，跟她一起參加海外的短期麵包課程，加深了我對麵包的知識，其後更一起到日本鳥越麵粉學校作短期深造。亦加深了我跟她之間的朋友關係。仙姐跟我可說是亦師亦友，沒有她作我的啟蒙老師，我應該是不會跟麵包結下不解之緣；但我卻有愧當仙姐的朋友，因為我記得她曾經説過，當朋友是需要兩肋插刀，她已為我做到，可惜我暫時卻未能做到。

她的新作其實都毋須多講，都是我們必買的，她每一本食譜都是實而不華，實用性高。

最後，在這裏祝願她的新作一紙風行，她的生活春風得意，我有機會為她兩肋插刀，成為她的真正朋友。

Louisa Kiang

香港 TOTUS 公司社長

鳥越麵粉學校八期同學

第一次認識仙姐是在台灣的麵包講習會上，當時已經知道她是香港有名氣的導師。但她並沒有任何架子，覺得她是一位樂於分享技術、容易相處的人。慢慢熟識她就會發現仙姐是一位十項全能的烘焙師傅，而且是一位不藏私非常大方的人。不僅如此，仙姐喜愛烘焙的程度，我會以「沉溺」來形容。為了使用最新鮮、天然的食材，仙姐更會親自耕種，此熱誠令我非常佩服。仙姐為這本書灌注了想不到的精神和心血，我熱烈推薦仙姐任何一本書本。

Vincent Lai

香港享樂烘焙創辦人

鳥越麵粉學校九期同學

在澳門從事教育工作的我，很希望藉著進修，把更多美味及專業知識帶給學生，奈何當年在澳門找麵包老師不容易。有天在書店看到獨角仙的《天然麵包香》，發現她的食譜入屋之餘又不失創意，又覺得這個名字很耳熟，馬上在網絡搜尋她，知道她正在開辦課程就立刻報名，就認識了仙姐這位好朋友。一連三個月，每逢星期三到香港上課，仙姐鼓勵我繼續學習，她還介紹我到台灣及日本研修；有些人會認為同行如敵國，反之，仙姐這位老師一點也不吝嗇，更望身邊的人與她一同進步，非常敬佩她有這樣的胸襟。

此書由淺入深，所講解都是基本知識與步驟，初學者也可以容易理解，在家亦能輕易烤出美味健康的麵包。

Rowena Lam

澳門嘗 ・ 手作烘焙教室創辦人

鳥越麵粉學校八期同學

相信好多人認識獨角仙都是由藍色大門開始，她可以說是烘焙 Blogger 界的鼻祖！

而我跟仙姐相識是由鳥越開始……

仙姐是一個毫無機心、願意提攜後輩的大師姐，她從不藏私，當她發現有關烘焙的新資訊時，她例必通知我們鳥越六人組。

仙姐熱愛烘焙程度相信比很多在職師傅更瘋狂……她經常到世界各地跟隨不同的名師學習，對烘焙的熱試多年來有增無減，使我不得不心悦誠服！

由《天然麵包香》開始到今日的《初嘗麵包香》，就如仙姐的本名一樣～堅！堅持不懈、堅定不移、堅苦卓絕、堅韌不拔……

陳浩然

香港嘉多利麵包店創辦人

鳥越麵粉學校十期同學

學習要抱着歸零的心。

造麵包有鑽研不盡的學問和技術，

以下是獨角仙歷年來在海外深造及拜訪麵包名店的留影。

自序

懂得造麵包已多年，由最初連麵粉都分不清高低筋，酵母亦不知怎使用，造出來的麵包像石頭，到上正式的專業麵包課程，甚至在日本的麵包店實習，這段長達廿多年的日子讓我深深愛上麵包 。

造麵包就像一本厚厚的書，有讀不完的學問和技術。記得當年寫 Blog 有一個願望，是帶動烘焙事業，讓更多人培養造麵包的興趣。事隔十年，看到現今在家造麵包的朋友也造得一手好包，甚至到海外上麵包課程，目的都是能自己造出美味的麵包給家人和愛人享用。這樣的熱潮，更隨着台灣在世界麵包比賽奪冠開始，直捲過來香港，各大教室都聘請台灣、日本甚至法國師傅過來授課，可見自家造麵包已是生活不可缺少的部份，所以這熱潮只會有增無減。

出過兩本麵包書後，發覺應該將基本及正確的造麵包概念和知識跟大家分享。台灣鐵能社社長林小姐在鳥越麵粉學校課程說過「學習要抱着歸零的心」，謙虛而積極，學習後融會貫通，亦要了解材料、造法和思考為甚麼要這樣做，加上多練習，就能成為民間中的高手，贏盡各人讚賞了。

獨角仙

鳥越麵粉學校八期生

目錄

材料介紹

原材料

麵粉

麵粉是烘焙製品中最基本同時也是用量最多之材料，選擇對的麵粉，對於造麵包來說是至為重要的。麵粉直接影響了成品的口感、組織和形態。麵粉由小麥磨製而成，小麥最主要的來源是美國、澳洲及加拿大，小部份來自中國、阿根廷、印度或日本等國家。小麥的種植由於品種、季節、地域、氣候等種種因素之不同，因而影響小麥品質之迴異，不同品質之小麥根據不同麵粉廠商的研磨技術會研磨出各種不同特性之麵粉，供應給各種不同需求之烘焙製品所使用，種類可以達到七十種。

從蛋白質的含量看，生長在大陸性乾旱氣候區的麥粒質硬而透明，含蛋白質較高，達 14-20%，麩質強而有彈性，適宜製作麵包；生長於潮濕條件下的麥粒質軟，麩質少，含蛋白質 8-10%，適合製作蛋糕、餅乾。

法國人對小麥粉的分類，基於小麥粉所含的「灰分」，大致分成 6 類。灰分指的就是小麥麩皮中所含的礦物質成分，也決定小麥風味的豐富程度。

根據灰分的含量，用數字的大小，來限定麵粉的形態，比如 T45 / T55 / T65 / T80 / T110 / T150。T 後面的數字越小，表示這類麵粉的精製程度越高，麵粉的顏色也越白。T 後面的數字越大，表示這類麵粉的精製程度越低，麵粉顏色也相對較深。法國小麥粉的分類，是以灰分的比例含量決定的，法國依照所含的灰分的含有量分成六種不同的形態 Type=T= 類形。

德國則稱為 Type 405、Type 550、Type 812、Type 1050 等等。

意大利小麥粉分類方式又與上述不同，分成 2、1、0、00，標準是麵粉研磨的細致程度以及移除多少麥麩和胚芽，以 Tipo 00、Tipo 0、Tipo 1、Tipo 2、Integrale 等來區分。

歐洲麵粉的筋性比日本麵粉來得高，搓揉過程比較吃力，不同季節、不同批次的歐洲麵粉，即使是同款的麵粉，麵粉的特質也會有很大的變化。

T45

- 灰分含量小於 0.50%
- 可以製作甜點、吐司和布里歐修等重牛油麵包
- 使用 T45 製作布里歐修的時候，相比用日本粉，水份需要減少
- 用 T45 麵粉製作的麵包會保留小麥香，不易被奶油搶走香味

T55 = Type 550 = Tipo 0

- 灰分的含量 0.50%-0.60%
- 可以製作法國麵包（老麵發酵法）和牛角包

T65

- 灰分的含量 0.62-0.75%
- 可以用於製作法國麵包
- 用 T65 製作的法棍，外皮酥脆，與 T55 相比，顏色更黃，皮厚酥脆

T70

- 灰分的含量 0.6-1%
- 特淡裸麥麵粉

T80 = Type 812 =Tipo 1

- 灰分的含量 0.75-0.90%
- 半粒粉，淡色全麥麵粉，保留部份麩皮

T85

- 灰分的含量 0.7-1.25%
- 淡色裸麥麵粉

T110 = Type 1050 = Tipo 2

- 灰分的含量 1-1.2%
- 全粒粉（大型法國麵包）

T150 = Type 1600= Tipo Integrale

- 灰分的含量 >1.40%
- 深色全粒粉（全麥麵包）

T130

- 灰分的含量 1.2-1.5%
- 深色裸麥麵粉

T170

- 灰分的含量 >1.5%
- 深黑色裸麥麵粉

麵粉的構成

小麥是由蛋白質、碳水化合物、脂肪、礦物質、鹽類、維他命及水所構成，其成分因產地、氣候、土壤及品種不同而異，最大的差別在於蛋白質之量與質。

蛋白質（Protein）

小麥中所含之蛋白質可分為麥穀蛋白（Glutenin）、醇溶蛋白（Gliadin）、酸溶蛋白（Mesonin）、白蛋白（Albumin）、球蛋白（Globulin）等五種。前三者不溶於水，後二者極易溶於水而流失。如果以水洗滌麵粉糰時，麥穀蛋白和醇溶蛋白互相黏聚在一起稱為「麵筋」，此兩種蛋白質與其他動物性或植物性蛋白質不同，最大的特點是會互相黏聚在一起成為麵筋，後二者就會流失在水中。

小麥中蛋白質之含量以麥芽最高，麩皮次之，胚乳最少。一般小麥之蛋白質含量，以玻璃質地之硬麥最高，粉質地之軟麥最低，硬質春麥與春紅麥蛋白質含量在 13.0-17.0%之間，普通各種軟麥則在 8.0-12.0%之間。

麵粉的類別與品質：依粗蛋白含量分為高筋、中筋及低筋三類

類別	高筋麵粉	中筋麵粉	低筋麵粉
水份	14.0%以下	14.0%以下	13.5%以下
粗蛋白質	13%以上	10-12%	7-9%
灰分	0.80%以下	0.55%以下	0.50%以下

高筋麵粉顏色較黃，質感輕爽和有光澤，用手抓不易黏成一團，主要使用於一般麵包製作。因各品牌不同而吸水量有差別，所以在轉換慣用的麵粉時務必把配方中的水份預留一些，以作調整。另外我們一般使用的手粉都是高筋麵粉。

中筋麵粉通常用於製作批餅、麵點。低筋麵粉用於製作蛋糕、餅乾。

碳水化合物（Carbohydrates）

小麥中碳水化合物可分成溶解性碳水化合物（Soluble carbohydrates）及纖維（Fiber）二種。

脂肪（Fat）

小麥中，以麥芽之含脂肪量最多。

灰分或礦物質鹽類（Ash or mineral salt）

將小麥或麵粉完全燃燒後的殘留物，絕大部份為礦物質鹽或稱為灰分，在小麥或麵粉中灰分含量是很少的。灰分的多寡，對麵包烘烤後的風味有更直接的影響，尤其對歐式麵包來説是主要的香氣來源。在相同的條件中（麵糰筋度、發酵時間、整型方式與烘焙溫度時間），麵粉中灰分含量較高的麵糰，所做出來的歐式麵包內部香氣才會更濃厚。

維他命（Vitamin）

小麥之胚芽含有豐富的維他命 E，小麥中尚含有少量的維他命 A 及微量之維他命 C。

水份

小麥之含水量在 13.0-15.0% 之間。

不同品牌的麵粉有不同的風味和特性，但初學者只需要能夠分別高筋麵粉、低筋麵粉便可以了。練習的時候不妨做點記錄，以相同麵粉使用不同配方比例食譜或以不同麵粉試同一食譜，便可以找出每種麵粉的特性和風味了。

酵母

酵母是由微生物培養出來，它為麵包賦予生命。

酵母分天然酵母和商業酵母，天然酵母取自天然食材例如：小麥、裸麥、米飯、水果或酸乳酪在適合溫度下培殖。而商業酵母是廠商使用薯仔、提子等基礎材料，在實驗室從微生物培殖出酵母菌來，將它大量工業化複製，產生出商業酵母。商業酵母含有大量的酵母菌，相比天然酵母，商業酵母比較穩定，保存期較長，購買方便。

酵母在發酵過程中，當中的酵素能夠分解麵糰中的蛋白質成為葡萄糖供酵母菌食糧，增加數量，並排出大量二氧化碳和酒精令麵糰膨脹，增加麵筋，形成結構鞏固和有彈性的麵糰。

酵母於 0-10℃進入冬眠狀態，發酵停止，於 20-32℃為適溫階段，50℃會自行消化，於 65℃以上就死亡。

市售的商業酵母可分為鮮酵母、乾酵母和速效酵母三種，而市售的天然酵母亦有分小麥、蘋果、提子、米等材料提煉出來，相比起自家釀製的天然酵母穩定。

不論哪種酵母，用不完的都應存放在雪櫃。

鮮酵母

含有大量水份，必須存放在低溫的環境。在0-5℃雪櫃可存放一個月，如存放在冷凍櫃雖然可以保存較久，但活力會大為減弱，用量亦要相應增加。鮮酵母濕度約有70%，發酵速度也較快，使用方便。本書建議使用鮮酵母，因它的風味最佳。

用法1：將麵粉及各種輔料放入大碗後，將鮮酵母直接搓碎均勻撒在麵粉上，充分拌勻後加水搓揉至麵筋形成。

用法2：將鮮酵母搓碎加入部份水中溶解，在搓揉操作中加水階段時加入，水溫根據氣溫而定，氣溫低水溫要稍高，水溫最高不能超過40℃。

用量：製作一般麵包，用麵粉量的2-3%，用量隨氣溫或製作品項調整，隨鮮酵母存放時間延長，相應要加大用量。

要知道鮮酵母是否新鮮，用手捏它是否硬實、有否彈性，不發軟、不黏手，外觀乳白、顏色統一，聞起來有酵母的清香味、無異味為鮮。如酵母顏色有異或味道轉變，表示酵母已經變壞，不能使用。

保存本地牌子鮮酵母的方法

將買回來的鮮酵母用密封膠袋袋好，用手把酵母捏至細碎，立即放冰格儲存，用時從冰格取出，可解凍，或添加份量內水份溶解使用。

但台灣、法國、意大利等地方的鮮酵母就未必能耐零下低溫，建議放雪櫃保存並儘快使用。

乾酵母

乾酵母是將鮮酵母壓榨成細小圓粒狀，再將其風乾，令其成休眠狀態，不易變質。乾酵母使用時須加入少量溫水和糖浸泡約10-15分鐘令其軟化和恢復活力，才可加入其他材料中攪拌使用，如果沒見到有泡沫就表示酵母失效了。它的特性是發酵耐力持久，酵母味道濃郁，但由於使用不方便，故不廣泛使用。乾酵母的用量是鮮酵母的一半。

速效酵母

速效酵母的製作和乾酵母相同，呈細小短條粉粒狀，可直接加入材料中攪拌，使用方便。如製作少量麵包，建議購買個別包裝的速效酵母，因酵母一經開封便會慢慢失去活力，這情況往往在你很用心地搓完麵糰後才會察覺。速效酵母用量是鮮酵母的三分之一。

市售天然酵母

熱門的品牌有星野天然酵母、白神天然酵母、LV1 等等。

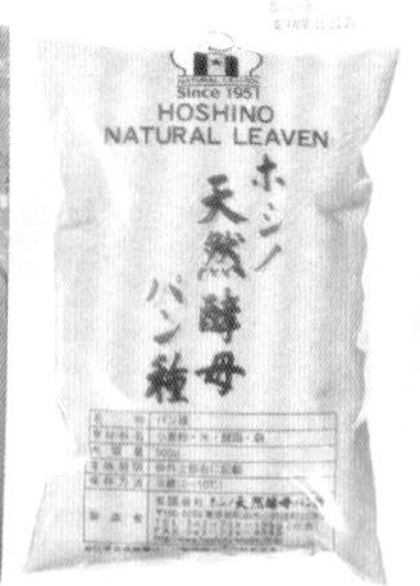

野生天然酵母

是自家培養的酵母，氣味較複雜，發酵力亦不穩定。

鹽

鹽是麵包製作的一個重要元素，它不單可增加麵包的香味，還能在麵包發酵過程中強化麵糰中的麩質，改善麵糰的柔韌性和彈性，增加麵糰的膨脹力。鹽也能夠控制酵母的發酵，也能殺死麵糰衍生的雜菌，沒有加入鹽的麵糰發酵較不穩定和過度發酵。但過量用鹽除了令麵包過鹹外，亦會給酵母壓力，阻礙發酵。鹽和酵母最好不要放在一起，鹽的使用由 0.5-3%，一般麵包使用 2%，因應鹽不同的種類而有不同的鹹度。

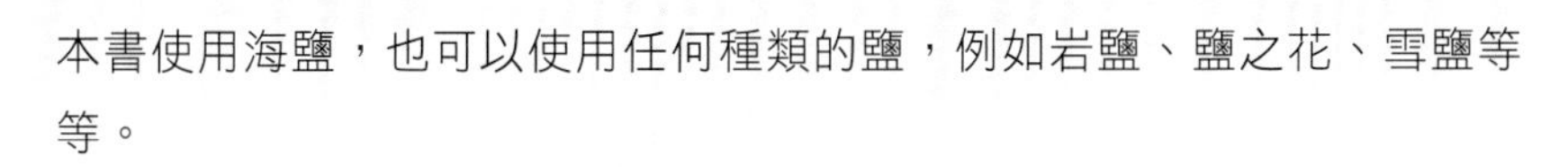

本書使用海鹽，也可以使用任何種類的鹽，例如岩鹽、鹽之花、雪鹽等等。

水

水份是造麵包的四大材料之一，水的用量可以參考麵粉商建議的吸水量來使用，一般麵粉的吸水量大約 65-70%，因應不同的麵粉牌子而有不同，也因應麵粉儲存和氣候的轉變而影響麵粉的濕度。食譜上的用水量只是參考，用手搓麵或用機搓麵都有不同用量，亦因應搓麵的熟習與否可作調整，初學者如果不習慣濕而黏手的麵糰，可以將用水減 3-5%，待熟習後可以慢慢加回，當然這樣麵包成品就會比較乾身；如果轉換麵粉時也要留意用水量，可預留少許用水，觀察麵糰情況去調節。食譜用的水份包括奶、果茸果汁、菜茸菜汁，如果以這些東西代替水都按水份計算。水有軟水、硬水之分，即是酸性水和鹼性水，造麵包一般使用 6.6pH 值的水。

副材料

糖

糖份是供給酵母養份的主要來源，除了增加風味、使麵包柔軟、增進色澤，更能保持成品濕潤，加強防腐。使用量大約在 0-30% 之間。適量的糖份會令酵母更活躍，但亦會對酵母產生滲透壓，壓力使酵母活力減弱；這時候有耐糖酵母的出現，亦可以將糖份分段放入，減低壓力。砂糖是常用的糖份，亦可以用蔗糖、紅糖、黑糖、蜂蜜、楓糖等等代替，來製作出更具風味的麵包。各種糖類甜度和風味各有不同，只要掌握它們的性質，就可自由搭配出不同的口味。

奶

奶可以增加麵包的風味、營養價值和色澤，並增加成品的柔韌性。新鮮牛奶的蛋白質中含有多量活潑性硫氫根，可使奶內乳清蛋白變性，會令攪拌後的麵糰減少吸水性，黏手並無法脹大。新鮮牛奶用來做麵包時必須先加熱至 85，使其失去硫氫根的活潑性，冷卻後再予以使用。除了新鮮牛奶，其他如奶粉、保鮮奶、淡奶等在加工時均已經過加溫處理，所以不會影響麵糰的性質。

我們不可將配方中的奶粉直接替換成等量的牛奶，因為奶粉的濃度大約是牛奶的 9 倍，1 份奶粉用 9 份水可以還原成牛奶，如果換成牛奶麵糰水份就會過多。奶粉用在麵包製作時，其最高用量最好不超過 6%；因為奶粉內含有高量的酪蛋白質，可增加麵粉內麵筋的強度，如果用量過高則麵包的組織將受到影響。

奶粉容易潮濕，在做麵包時奶粉可與砂糖拌勻，並要立即攪拌或搓揉，否則奶粉將吸取麵粉中的水份凝結成塊，無法再溶解於水中，影響到麵包的組織。因為奶粉屬於乾性原料，在使用時配方的水份必須隨着奶粉的用量來增減。麵粉中加入的奶粉量越多，表皮的顏色就越深，對應的烘焙時間也要控制得相對短一些。除了表皮之外，奶粉作為一種 pH 緩衝劑，還能保障麵糰正常發酵。

乳製品對麵包的另一個重要影響是可以提高麵包的抗老化性，也就是延遲腐敗。一般來講，麵包老化的主要原因包括澱粉的回生和水份的損失。比如，麵包硬度增加、易掉渣、風味變劣、芳香消失等，加入奶粉可以增加麵包的吸水率，減少水份流失，延長儲存時間。另外，例如法棍，就是因為配方中不用牛奶而用水，所以口感較硬。

脫脂奶粉、牛奶、鮮忌廉、酸乳酪、淡奶、煉奶、法式忌廉、軟芝士等等普遍使用於麵包製作中。

蛋

蛋可增添麵包的香味和提高食味，令麵包有焦皮薄、光澤、質地鬆軟。蛋脂肪中的卵磷脂，為水和油的中介，可當作天然乳化劑，並幫助成品保持濕潤。

油脂

油脂給予麵糰延展性，彈性更好，令成品更鬆軟，更添香氣和顏色，也能延遲麵包的老化。油脂可改善麵包的品質，在麵糰發酵時發揮潤滑作用，促進麵包體膨脹，令麵包鬆軟，延長成品保存時間，增加香氣和營養價值。一般麵包的油脂使用量約在 6-12%，Brioche 和 Danish 除外。油脂過多令麵包內部組織粗糙，並延緩發酵速度。本書使用的油脂是無鹽牛油和橄欖油。

常用的油脂有固體油如牛油、馬芝林（即植脂牛油）、酥油和豬油，液體油如橄欖油、米糠油等等 。

液體油通常和材料一起搓揉或攪拌，而固體油則視乎麵糰的狀況和自己希望的口感而投放，一般會在搓揉或攪拌中期，讓麵粉吸收了水份後投油，又或者等麵糰的麵筋組織成形後投入，早一點投油，麵包咬斷性會較好（麵包很容易被咬斷的意思）。點心麵包、甜麵包的投油時間可以早一點；相反，待筋膜成形後投油，麵包就要出點力才能咬斷，例如吐司等比較紮實的麵包就可以待筋膜成形後才投油。

添加材料

我們可以把喜歡的材料加進麵糰，加入方法可以是搓揉入麵糰，可以用切割方式加入麵糰，也可以用包捲方式包入麵糰。

果仁

加入麵糰的果仁都需要預先用低溫焗爐烘香才能保持香脆，一般做法是預熱焗爐至 120-130℃後，把果仁放入焗爐焗約 10 分鐘，不需要上色便可以；出爐後放涼使用，不可以將熱果仁加進麵糰影響麵糰發酵。合桃、夏威夷果仁、黑白芝麻、榛子粒、杏仁、花生、南瓜子、葵花子及腰果等都是常用的堅果仁。

乾果

要視乎乾果的濕潤程度處理，如果夠濕潤的乾果，例如無花果可以直接使用；但如乾果很乾燥會吸去麵糰水份，在麵糰表面的乾果亦會被焗至過乾和苦澀，所以使用時可以將乾果泡水或酒類以增加濕潤和香氣。要記住不能過濕，在加入麵糰時要隔清多餘水份和徹底放涼；避免使用鮮果加入麵糰，這會令麵包過濕。提子乾、藍莓乾、草莓乾、無花果乾、蔓越莓乾都是常用的乾果。

堅果粒及果乾其實不太會在配方中起作用而影響麵糰本身口感，每種堅果及果乾的口感都不同，添加比例可隨自己喜好做調整；但用量也不可以過多，否則果粒會阻斷麵糰間的連結，容易造成烤好的麵包破碎。

芝士

包在麵糰裏的芝士通常使用半硬芝士，例如 Emmental、Cheddar、Gouda 等等，放在面的芝士可使用硬質芝士，例如 Parmesan 或可以溶的 Mozzarella 等。

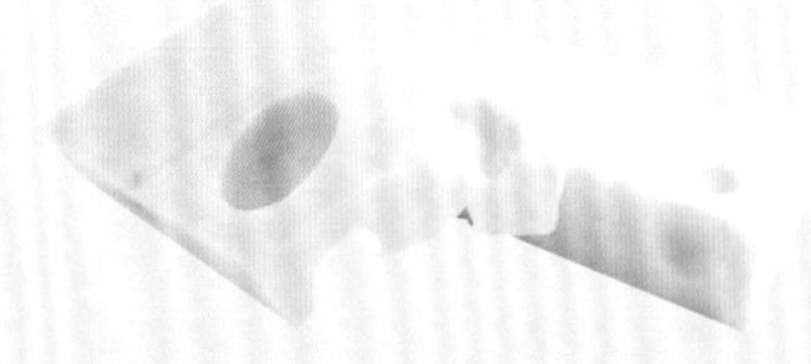

香料

可以加入鮮香草或乾香草、香料來增加風味，但要注意份量，不可過多。

朱古力

可以用粒狀的朱古力混合在麵糰內，或以片狀、條狀包在麵糰裏；粒狀朱古力亦最好使用能入爐的，以免朱古力太快溶掉影響美觀。

可可粉、抹茶粉及即溶咖啡粉

可可粉、抹茶粉及咖啡粉是最常見加入麵糰的材料。可可粉的使用量約為麵粉總量之5-8%左右，抹茶粉及咖啡粉大約在 3-6%，要替換的多寡當然沒有一定標準，但建議控制在 10%以內，要注意粉類的吸水性；所以隨着粉類用量增加，會明顯感受到麵糰越乾硬，要調節水量了。

抹茶粉和即溶咖啡粉品質差異很大，有分天然和加工兩種，有很大的風味差異，有些咖啡粉極苦，粉末顏色也有很大不同，所以建議從麵糰總重的 3%開始添加。

各式天然風味粉

市面上其實有許多天然風味粉可以使用，讀者可作多方嘗試，例如：蔬菜類的菠菜粉、南瓜粉、芋頭粉及牛蒡粉等；水果類的草莓粉、紅莓粉、蔓越莓粉、檸檬粉及柚子粉等，還有墨魚粉或海苔粉等，這類原料的用量和可可粉相同，除了要選擇天然製品之外，最好選擇味道重一點的原料，造出來的成品風味效果會比較明顯。

很多食譜中也會使用自己打的南瓜茸、菠菜茸、粟米茸，這其實也是一種改變風味的方法，但是自製果茸或蔬菜茸含水量較高，且水份含量控制不易。若要添加至味道能夠完全被凸顯的狀態，添加量就必須提高；問題在於麵糰配方的平衡會變困難，例如澱粉含量高的材料多下了會令麵糰變得無彈性。容易有失敗風險，所以要特別注意控制。

鮮果

可以放在生麵糰上面烤焗，亦可待麵包烤好後放上面使用。

造麵包的工具

工作台

在我們製作麵包的過程中，把所有配料在工作台上混合好是很重要的一個步驟。工作台最好高度適中，以身體彎腰雙手按住枱面計，工作台面到達肚臍或以下最理想。搓麵時我們要使用雙手壓下的身體力量，不要只用雙臂力，單使用臂力很容易弄傷手肌肉或令筋腱受損。另外，我們最好選用木質台面揉麵，麵糰不易黏在木質桌面上，而且木質傳熱及導熱慢，不會影響麵糰溫度；可是香港天氣潮濕，要保養木枱面是很困難的事，所以石質枱面是次理想的選擇。要挑選密度夠高，不易藏污納垢的麻質石材，亦因為較其他石材堅硬，不容易被刮花，而且溫度清涼不易傳熱，不會容易令麵糰升溫。仿石、強化玻璃、搓揉板等等都可以揉麵，但不是最好的選擇。如果選用的是薄身工作台，無論木材或石材，可能會和枱面咬不緊而滑動，可以用濕布放到工作台和桌面之間，這樣就可以固定工作台了。

電子磅

要準確量好各種材料就應使用精確的儀器，電子磅是最好的選擇，選購可以量 0.5 克起至 3 公斤的磅秤。

大碗

幾乎所有質地的大碗都可以用來拌勻材料。

切刀

幫助切割分開麵糰和清理枱面用。

膠刮

幫助清理攪拌用大碗或刮起麵糰用。

焗盤

烤焗麵包用，有時會配合焗爐石或厚身鋁片使用。

焗爐脫模噴油

可以噴在周轉箱或枱面防止麵糰黏着。

發酵盒（周轉箱）

我們使用適合麵糰大小的發酵盒，有助發展更好的彈性麵糰，使用導熱慢的容器，例如有蓋膠盒是最合適的，因為金屬容器會受室溫影響而影響麵糰發酵的溫度，而玻璃器皿較容易打爛；所以塑膠盒是最好的選擇，蓋上蓋子可以防止麵糰乾燥龜裂。

不沾布、矽膠蓆或牛油紙

墊在焗盤上防黏，清洗容易。牛油紙、不沾布較矽膠蓆薄，用來焗歐洲包較易傳熱。

焗爐

市面有用氣體焗爐，也有用電的焗爐。氣體焗爐只有上火、下火，有不足之處，所以我不太喜歡使用氣體爐。電焗爐有坐枱式、有嵌入式、亦有連電熱板的組合爐；有只行熱風的、有可調校上下發熱線的、亦有可以熱風加發熱線一起用的。我們需要知道所用焗爐的性能，由最基本的使用多少電力、可以同時間焗多少層麵包、焗盤大小、焗爐門有沒有膠邊可封密熱氣、安放位置是否適合等等都是需要關注的。接着是使用方法，當我們使用熱風焗麵包時，要注意因為熱風對流是焗爐內裝有一把風扇，風扇周圍有發熱線，風扇開動將發熱線的熱力運行到爐的每個角落，成品色澤均勻，恆溫也好；可是麵糰初入爐被熱風吹着，麵糰表面會很快糊化結焦，但這時候的麵糰尚未發酵到頂點，令到麵包體積細小，亦會變形。所以，我們使用熱風功能時，先將焗爐預熱得比烤焗溫度高 40-50℃，入麵糰後就把焗爐關機 2 分鐘，使麵糰不會易結焦而影響表皮，麵糰得以充分膨脹。只用發熱線的就沒有這顧慮。

麵包製作的

十二個基本步驟

計算並量度材料

計算好所需材料份量，使用精確的儀器，量好各種材料。

混合、搓揉

把麵粉、水、酵母、糖、鹽等等材料混合，搓揉出筋性，直至麵糰表面光滑和有彈性，使酵母平均分佈於整個麵糰中。材料預備好我們就需要開始搓麵糰，首先將材料混合 ，通常配方都會要求我們把酵母溶於水，然後就是加入麵粉，再加入其他配料（如鹽和砂糖），最後再加牛油。

用手搓麵糰或機打麵糰時，都要以慢速混合材料，這是非常重要的一個環節。水會很快就被麵糰吸收，並形成鬆散的濕麵糰，便可以將麵糰移至工作台繼續搓揉了。

＊步驟圖請看「圓形小餐包」P.38(圖 1 - 16)

發酵

將麵糰放在大碗或大盒內，置於室溫下（約 23-30℃，冬天用暖水保溫）進行發酵。當發酵至約兩倍大，可測試麵糰。食指沾少許麵粉，慢慢戳入麵糰中，如麵糰發酵適當，指孔不會收縮；如指孔迅速回縮，即發酵不足；若戳下時整個麵糰洩氣、收縮，便是發酵過度了。

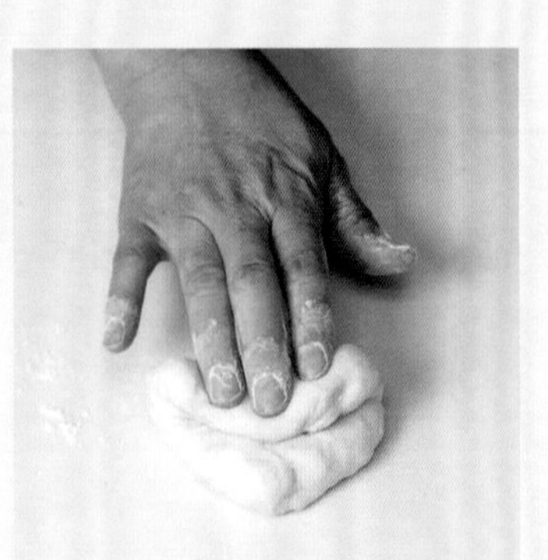

＊步驟圖看「圓形小餐包」P.40(圖 20 - 27)

排氣

用手輕輕拍打麵糰排出空氣，或取出麵糰，把麵糰邊緣摺向中央。此動作可鬆弛筋膜，並平衡麵糰溫度，令麵糰發展得更好。

分割

用刮刀和磅秤平均地分割麵糰。動作要快，否則各個小麵糰會發酵不平均。

滾圓或輕捲成條狀（Handsquare）

分割後將麵糰滾成圓形或捲緊成條狀（Handsquare），伸展麵糰至外表光滑，使完成品更有吸引力、更有光澤。當製作圓形、欖形、三角形麵包的時候我們做滾圓；當做辮子、捲吐司或長條形麵包時我們用 Handsquare。

* 步驟圖看「圓形小餐包」P.41（圖 32 - 61）

延伸發酵

麵糰滾圓或輕捲成條狀後，放在桌面或周轉箱內，蓋上保鮮紙或蓋子，作伸延發酵，大約15-20 分鐘，讓緊縮的麵糰鬆弛，易於造型。如特別需要，滾好的麵糰會放 0-5℃雪櫃延伸發酵，這樣會令麵糰更容易處理，如白汁雞包（看 P.124）。

造型

按需要做成不同形狀，如用模具，要先塗油防黏，方把麵糰放入，留意麵糰收口朝下；使用藤製麵包籃宜先篩上麵粉，收口則朝上。

* 最後發酵

把麵糰放在溫暖地方發酵至適當體積。

* 麵糰經過造型後便進入最後發酵，因應麵糰大小，食譜材料不同，室內溫度不同，氣溫不同，發酵時間都會不一樣。甚至於相同麵糰、相同溫度，每天的發酵時間也會有變。我們要累積經驗，以觀察、觸摸來判斷麵糰完成發酵沒有。適當的溫度大約在 28-32℃，並不是説越暖越好，要讓麵糰有充足的時間發酵，才可以造出好麵包。過熱、過冷、過乾燥或過濕，都會影響成品。過熱麵糰會發酵過頭，內裏組織粗糙，酒精味強烈，成品會乾硬，凹陷；過冷麵糰會發不起來，成品不夠鬆軟，體積細小；乾燥會令表面龜裂，影響外觀，要常常幫麵糰保濕；但過濕就會令麵糰變形，影響外觀。一般小餐包最後發酵大約 30-45 分鐘，大型麵包大約 45-90 分鐘，吐司麵包大約 50-90 分鐘 ，角食吐司麵包要看準時機才入爐，發酵時間就要掌握得很好。記着初學者要預留多一點時間預備配合上光、飾面時間和預熱焗爐啊！

烤焗

麵糰最後發酵完成，可作最後裝飾（如塗蛋、篩麵粉、灑芝麻等等裝飾包面、用不同模格上粉或㓟上或剪出花紋等等），之後放入已預熱焗爐烘烤。每個焗爐的火力和操作都不同，所以要熟悉自己的焗爐運作以作調校。

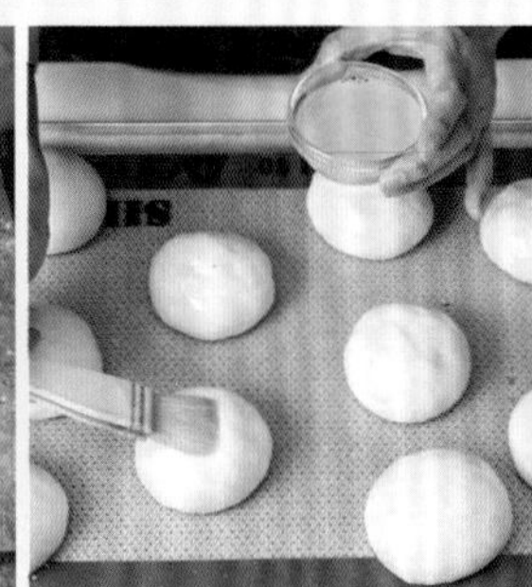

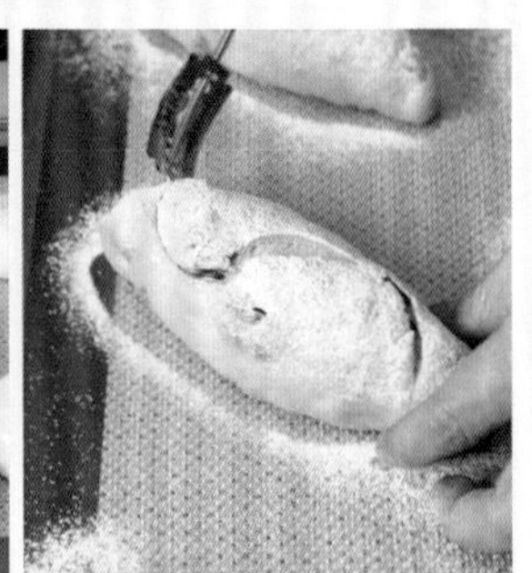

冷卻

如使用模具烘烤，出爐後要立即脫模，並放在網架上冷卻，以免水蒸氣弄濕成品。出爐麵包內部充滿水氣，應待麵包冷卻或只有些微餘溫時才可切開或食用，否則麵包組織會黏在一起，變得糊口。

儲存

麵包由出爐一刻開始老化，盡量新鮮食用。家庭製作的麵包不加入添加劑，一般比市售麵包容易變得乾硬；沒有餡料的麵包可在室溫存放約三天，有餡料的就要立即放雪櫃保存，用保鮮膜把麵包包好或用膠袋載住封口，可在雪櫃保存 7 天左右；如果要保存得更久，要貯存於冰格冷凍，可保存多月，食用前取出置於室溫解凍。翻熱麵包只要放焗爐以 130-140℃低溫，噴水在麵包上焗熱數分鐘即可回復新鮮可口；或用微波爐、電飯煲翻熱亦可以。

麵包製作的
基礎理論

烘焙百分比

烘焙百分比是烘焙產品製作配方材料所設定的一種計算方法，此方法是根據麵粉的重量來推算其他材料所佔的比率；換句話說，不管配方中的麵粉重多少，我們也設定為 100%，而其他材料的重量是以各佔麵粉的百分之幾來計算。

公式：烘焙 % =（材料重量 ÷ 麵粉重量）x 100%

材料	重量（克）	烘焙百分比（%）
麵粉	1000	100
糖	150	15 = (150÷1000x100)
蛋	100	10 = (100÷1000x100)
鹽	20	2 = (20 ÷1000x100)
酵母	30	3 = (30 ÷1000x100)

烘焙百分比換算為各項材料的重量

公式：重量 =（麵粉重量 x 材料烘焙 100%）÷100

材料	烘焙百分比（%）	重量（克）
麵粉	100	1000
糖	15	150= (1000x15)÷100
蛋	10	100= (1000x10)÷100
鹽	2	20= (1000x2)÷100
酵母	3	30 = (1000x3) ÷100

獨角仙有一個更快的方法，就是心目中定下你想做的麵粉份量，例如 500 克，那麼你就將所有份量乘以 5，就是我的烘焙百分比了。

而計算所需材料算式，是將願望麵糰重量除以百分比總和了。

公式：份量 = 需要麵糰重量 ÷ 百分比總和

例如：假設我要造 10 個 60 克的麵包，即是我需要 600 克麵糰，以 600 克 ÷ 百分比總和就知道所需材料的百分比數目是 3，就可以列出以下份量了：

份量 = 600 克 ÷ 200% = 3

材料	烘焙百分比（%）		份量		重量（克）
麵粉	100	x	3	=	300
糖	15	x	3	=	45
蛋	10	x	3	=	30
鹽	2	x	3	=	6
酵母	3	x	3	=	9
水	70	x	3	=	210
總和	**200**				**600**

水溫計法

很多時候造麵包失敗的原因是麵糰溫度管理得不好，搓揉好的麵糰，無論手搓或機搓都應維持在 24-26℃室溫。材料、溫度都會影響麵糰溫度，在搓揉麵糰時，要留意室溫，可以用空調去控制室溫。理想的室溫大約在 23-26℃。材料溫度也可以控制，例如把它們預先放雪櫃雪凍。

控制用水溫度是最理想的辦法，通常是夏天用較冷的水，冬天可以使用較溫暖的水，水的溫度是可以計量出來的。

但不是越冷的水就最好，因為麵粉接觸低於 20℃的水，麵筋很容易收縮，不能搓出鬆緊有致的麵糰，酵母亦會因過冷而影響發酵力。所以水溫是要注意的一環，不耐低溫的酵母可以在麵糰搓至超過 20℃才下就成了。

麵糰願望溫度 x 3 -（室溫＋粉溫＋磨擦溫度）= 水溫

這條算式是給麵包店用的，我們在家造麵包要留意用手搓揉麵糰和機搓是不同的，手的力相較機器比較不會容易升溫，在搓揉時磨擦所產生的溫度不會很高，尤其是手比較涼的朋友，搓揉 15 分鐘可能只升兩度。因應不同打麵機，磨擦力都有不同，越大馬力的機器，在搓揉麵糰升溫越慢，越多麵糰亦會令機器發熱得更快；所以在用手搓揉麵糰時所用的水溫大約比室溫少 2-3℃，在 22-24℃便可以，如在冬季，水溫可以用 24-26℃ 。

如果機器打麵，一分鐘麵糰可以升溫 0.8-1℃，如果打 10 分鐘麵糰，則會升 8-10℃，以算式來計：

26 x 3 –（25 + 25 + 10）= 18℃

烘焙的基本

麵糰造形後，我們進入最後階段就是烤焗，當麵糰整形時，我們要預熱焗爐，因為開爐到爐溫足夠可能需要 15 至 30 分鐘時間，麵糰如果放入未經預熱的焗爐會過度發酵，過度發酵的麵糰，內裏的酒精會把水份蒸發很多，因而令麵包乾硬。如果二氧化碳氣體太多，亦會撐得麵糰表面過薄，容易爆破，失去彈性，導致麵包扁塌、皺皮。

焗爐預熱的時候，預熱要把爐溫比烤焗爐溫高 20-30℃，因為一般家用焗爐體積不大，恆溫亦不會造得很好，當放麵糰入爐時要將爐門打開，當爐門打開時爐溫會下降 20-30℃。你把麵糰放入後要等很久才能升溫至所需爐溫，這樣麵糰就會花多很多時間才能成形及上色，這樣會導致麵包表面乾硬和色澤不漂亮；把爐溫調高後放入麵糰，再調校所需溫度，烤焗就會更有效率。

很多同學問我需要買一個焗爐用溫度計嗎？我覺得焗爐內每個角落的溫度也有不同，溫度計只能探測某個角落的爐溫，它不能幫你令爐溫平均；所以如果你在焗麵包時發覺麵糰有部份不易上色，可能是面色、底色或者是側面，那就是焗爐溫度不平均，這樣你便要把盛麵糰的焗盤調位，來相就爐火，才能解決爐火不平均的問題。

造麵包的方法

現今製作麵包分很多種方法，以下略作介紹：

直接法

顧名思義，直接法是將全部材料一次搓揉或攪拌成麵糰，在常溫經短時間發酵後製作麵包。優點為發酵時間短，製作流程簡單，製作的時間短，麵包的風味可以直接展現出來；缺點是麵糰含水量低，老化較快，麵糰欠缺彈性，也沒有充分發酵的芬芳。亦有將直接法麵糰放入 0-5℃雪櫃低溫發酵，麵糰經 24-72 小時長時間發酵，麵糰充滿香氣和風味。

書內採用直接法的麵包：
圓形小餐包（P.36）、軟小餐包 - 編織包 3、5、8 號花（P.66）、朱古力吉士包（P.130）

直接法 — 翻面

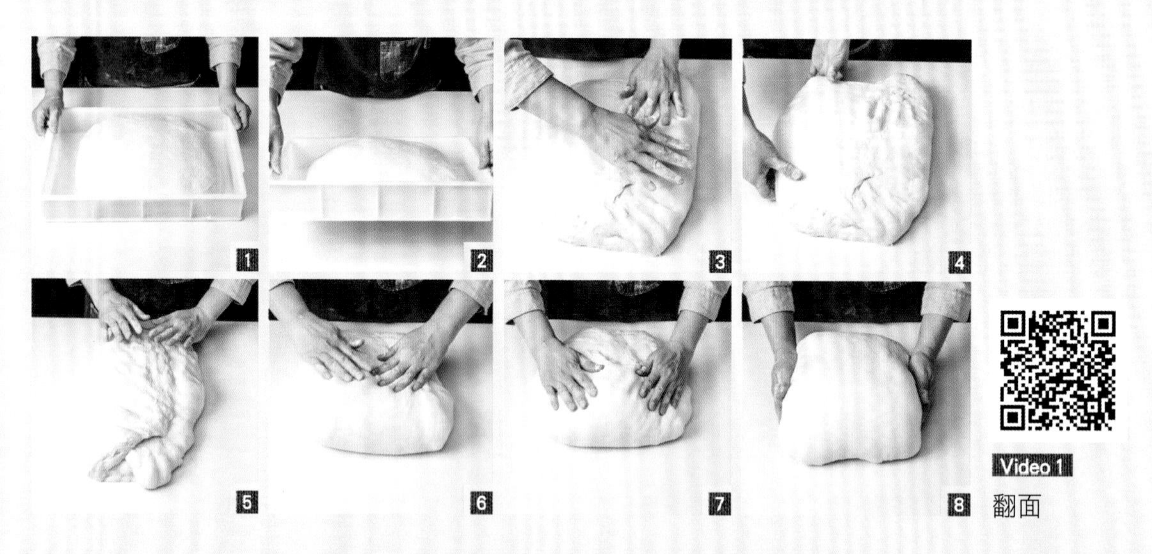

翻面

是直接法一種，需要較長時間的發酵讓麵包得以增加風味和彈性。當麵糰在一般情況下到適當時間，麵筋就會膨脹到達頂點，麵糰內裏溫度上升，酒精和醋酸菌開始增多至麵筋不能承受階段，這時我們可以利用翻面來平衡麵糰溫度，並能夠增加麵糰延展性和讓酵母數量增加。翻面可以造出有彈性及有嚼勁的麵包，比只用直接法好。

書內採用直接法 - 翻面的麵包：
全麥包（P.72）、蔬菜包（P.102）、提子包（P.108）、白汁雞包（P.124）、粟米沙律包（P.162）

中種法

中種法是將部份的粉類、水、酵母預先製作，讓其經過適當時間的發酵熟成，再加入其餘的粉類及其餘材料繼續搓揉或攪拌成麵糰再發酵製作成麵包。中種法亦有很多種類，例如：50% 中種法、70% 中種法、80% 中種法、有糖中種法、常溫中種法等等。

中種法又稱二次發酵法，是將麵糰分兩次攪拌，先攪拌中種麵糰，使其經過一段時間發酵，再與其他部分混合攪拌形成製作麵包的麵糰。因為酵母最怕高糖量、鹽和油脂，中種法能使酵母在第一次麵種中發酵良好，使水份完全滲透於麵粉，可讓麵糰水含量增加，經過第二次攪拌後，把氣泡全打掉，達到柔軟、組織細密效果。

用中種法製作麵包，麵種的溫度環境非常重要，發酵時間過多或不足都不能製作好吃的麵包，時間溫度是最關鍵。中種法麵糰最後醒發較快，體積大、柔軟、老化慢、是比較容易操作的方法。以 70% 中種比較受歡迎。70% 中種是指中種的麵粉和主麵糰的麵粉比例為 7：3，加入鹽、酵母和水製作中種後，放常溫或低溫發酵後再加入主麵糰，混合攪拌成麵糰製作麵包。

有糖中種法，是將部份糖類、麵粉、酵母和水先造成中種，再和其餘材料搓成麵糰，主要用於糖份含量高的點心麵包。

書內採用中種法的麵包：
椰茸包（P.88）

老麵種、發酵種法或法國老麵法

這些是已經發酵過一次的麵糰，把它添加於麵糰中，讓麵糰在長時間的發酵中讓麥香完全釋放，造就出麵包散發出小麥天然香氣，並使麵包口感更煙韌、更鬆軟香醇、組織更細緻。很多研究指出把酵母放進麵糰需要 45 分鐘的適應期，才會大量的釋放出二氧化碳和酒精，使用老麵法除了可以避開抑制酵母的條件（例如鹽和高比例的糖），更重要的是在前置發酵的期間，酵母跨越了 45 分鐘的適應期，讓麵糰的酵母量更快增加，令麵糰發展得更好。所以使用老麵和直接法最大的差異就在這裏。

老麵

老麵是很好的天然添加劑，可使麵包更有口感，保水力更佳，用量大約麵粉量的 15-30%。我們可以預先把老麵做好（即使含少量油糖也可），分割成約 50 克一份，冷凍儲存，做麵包前半小時放在室溫或溫暖環境下解凍即可，甚至用膠袋包好放進暖水中解凍。幾乎做甚麼麵包都可以加入老麵進麵糰內同搓。如非做大量麵包，也毋須增加其他材料份量，如用手搓麵包，在搓包前把老麵撕成小塊，放在麵糰份量內的水中稍浸軟，便更易融合。

老麵材料	百分比（%）	重量（克）
高筋麵粉	100	250 克
水	60	150 克
鮮酵母	0.5	1 克
鹽	0.4	1 克

做法

將以上材料搓成麵糰後在室溫發酵至少 4 小時，或者把它放在雪櫃任它發酵十多小時，取所需份量使用。餘下的可以分成每次所需重量，滾圓，冷凍保存約 2-3 個月，用時預先拿出來室溫解凍即可。（圖 1 - 8）

書內採用老麵種法的麵包：

大窿芝士餡包／吞拿魚沙律餡包（p.114）、紫薯花花包（p.136）

法國老麵法或發酵種法

法國老麵和老麵同樣都是經過發酵的麵糰，在百分比中只佔 15-30%，和中種比較，老麵是添加在麵糰的發酵種，對主麵糰的其他份量沒有影響，抽取出來也可以。而中種的材料和主麵糰是有份量的，不可亂作加減。法國老麵使用了法包專用麵粉和有用麥芽精，專門針對歐洲麵包使用，法國老麵可以替代老麵用於軟麵包，但老麵不能代替法國老麵使用在歐洲麵包。

搓好的麵糰大約是 24℃，放室溫發酵 90 分鐘後，放 0-5℃雪櫃 6-8 小時或搓好放 0-5℃雪櫃 12-24 小時。

發酵種材料	百分比（%）	重量（克）
法包粉	100%	100 克
海鹽	1%	1 克
麥芽精	0.5%	0.5 克
鮮酵母	0.5%	0.5 克
水	65%	65 克

書內採用法國老麵法或發酵種法的麵包：

硬小餐包（p.148）

湯種法或熨種法

「湯種」在日語裏意為溫熱的麵種或稀的麵種。「湯」的意思有開水、熱水、泡溫泉之意。造法是將部份麵粉加水混合後加熱至特定溫度，或將特定溫度熱水加入麵粉中，熱水使麵粉中的澱粉糊化，此糊化的麵糊稱為湯種。湯種再加其餘材料經搓揉或攪拌而成湯種麵糰。湯種麵包與其他麵包最大的差別在於澱粉經糊化後，能夠提高麵糰吸水量，使麵包氣泡細化，造出的麵包組織更柔軟，口感 Q 彈，保濕性強，可延緩老化。流行有 65℃湯種、85℃湯種等等。

書內採用湯種法或熨種法的麵包：

火腿芝士洋葱吐司（p.82）、腸仔包（p.142）

液種法

Poolish 是源自波蘭再傳到法國，是含水量較高的發酵種，製作配方要求是以 100% 的水和 100% 麵粉混合，再加入少量的酵母在常溫或 0-5℃發酵 12-24 小時，讓酵母有更多的時間對麵糰中澱粉和蛋白質充分發揮作用，令麵糰更具伸延性，減少搓揉出光滑麵糰的時間，延長麵包保存時間，保濕性佳以及創造更豐富的味道。常溫或低溫發酵的 Poolish 分別在於：常溫風味甘甜而低溫風味較有香氣。

Poolish 的成熟與否，可影響最終成品的風味，成熟的 Poolish 可令成品有着蜂巢般的組織，乳白的顏色，健壯渾圓的外表和漂亮的裂口，濃厚的麥味。現在更有利用水果汁或果茸來製作 Poolish，成品帶有果香和顏色，是很有趣味的製作方法，唯有一些水果例如木瓜、菠蘿和奇異果等水果因為酵素太高，不利發酵。

書內採用液種法的麵包：

紅莓合桃包（p.94）

低溫冷藏法

低溫冷藏法是將 50% 以上的麵糰預先搓好存放在 0-5℃雪櫃發酵 17-72 小時。

主麵糰的發酵時間相應地減少，最後發酵後勁凌厲。這種方法的成品，特質是口感有軟綿、濕潤，帶有發酵的芬芳，可存放較長時間而不易變硬，保濕良好。

可參考獨角仙另一本著作《天然麵包香》。

酸種法

酸種法是用裸麥粉加水搓成麵糰放置 4-5 天成初種，再用初種加麵粉加水續種，重複 3 次成酸種，加入主麵糰材料搓成麵糰製成麵包。麵包因應地方不同有不同的酸味和風味，主要流行於德國，美國舊金山的酸種麵包也很有名。

酒種

酒種是將熟米飯加釀酒的米麴發酵成酒種，加入麵糰製成麵包，成熟的酒種充滿米酒的芳醇，日本著名的麵包店木村家的酒種豆沙包已有百多年歷史，是人們不能忘懷的味道。

酸乳酪種

酸乳酪種是以酸乳酪加麵粉及少量酵母起種發酵而成，麵包有着乳酸芬芳。

水果天然酵母種

水果天然酵母種是以水果酵液起種，加入小麥粉發酵而成，是酵母菌及乳酸菌共存的酵母，只是起種用的材料不同，而有不同的菌種、不同的風味。

魯邦種

魯邦種是藉由天然酵母液（Starter、Mother、Chef 或起種）、裸麥粉或小麥粉加入水和粉續養而成的麵種。主要不依靠蔬果類的酵母菌進行成長，而是依靠穀物吸取養份成長。使用不同的麵粉，麵粉擁有不同的蛋白質和灰分，就會產生略微的差異。大家可按照自己的喜好，製作出清淡或者酸度明顯的麵包。

魯邦種分魯邦液種及魯邦硬種，魯邦液種以小麥粉、裸麥粉、水及少許鹽經發酵製成的天然酵種。魯邦硬種是最初的製作魯邦種的方式，過程很緩慢，風味及發酵能力會因為續種而流失，因為當時沒有雪櫃，人們就靠着每天重新攪拌來培養麵糰。現在魯邦種多以液態來培養，大多數情況下説到魯邦種都是指的魯邦液種，它的發酵速度較快，而且有了雪櫃能保存，操作上更加方便。魯邦種是酵母菌及乳酸菌共存的，乳酸菌還能增添麵包乳酸發酵的風味，它形成的酸味及發酵味，使麵包的味道更有深度，凸顯穀物本身自然的風味。

用機器打麵糰

要注意之處

除了手搓麵包，廚師機、麵包機或座枱式攪拌機都是可以用來打麵糰的，以下介紹各種用機器打麵糰需要注意之處。

打麵機

打麵機的機械力量可以做到人手達不到的效果，尤其是軟麵包，它可以打出細緻柔滑的麵糰。市面上的打麵機有很多型號和功能可以選擇，有些攪拌缸夠大夠深，有些顏色漂亮，有些功能眾多。如果造麵包的次數不多的話，市面上大部份的打麵機都能夠使用；但如果經常造麵包而且份量不小的話，一部馬力大，設計簡單的打麵機就會比較理想。記着，越多功能的電器越容易壞的道理，而且打麵糰是需要很大力量的，發動機會因為操作而發熱，熱力會透過攪拌勾傳到麵糰，這就是計算水溫算式所提到的磨擦溫度，麵糰會因此過熱（超過 28℃），要解決麵糰過熱的問題，我們要熟悉自己的打麵機的升溫狀況來調節用水溫度，才能管理好麵糰溫度。

座枱式攪拌機或廚師機

一般坐枱式攪拌機或廚師機打 200-300 克麵粉的麵糰是沒有問題的；如果要超過了，最好分兩次攪拌了。分兩次攪拌就要待機件降溫後才使用。

操作要注意的，是因為機動的磨擦力會令麵糰溫度升溫得較手搓快，平均機動攪打一分鐘會升溫 0.8-1℃，所以管理麵糰溫度要留意磨擦溫度，以用水或液體的冷暖來調節麵糰溫度。很多時候，麵糰未打出筋膜就已經到達 26℃，我們可以在碗外敷冰來降溫；但最好把這些情況做點記錄，積累這些經驗，就很容易掌握打麵糰的溫度了。

注意：麵包麵糰的攪拌千萬不要使用手提攪拌機，因為麵包麵糰一般較大較重，手提攪拌機的功率不足，會很容易燒壞裏面的發動機。

亦最好選擇當地有維修服務的機器。

麵包機

麵包機可以代替打麵機來攪打麵糰，但大部份的麵包機已設定了攪拌和發酵功能，但設定的自動程式，攪打時間都不足，無法達到麵包所需要的完全擴展麵筋階段，這就需要我們自己增加一下攪打時間，可以在攪打完成後關機，然後再開機攪打第二次，一般增加一次時間就夠了，具體要看麵糰的情況而定。或者可以選擇手動程式，自己掌握攪打、發酵和烘烤的時間。

烤焗麵糰的方法

直接烤焗

麵糰發酵好後，把麵糰轉移到裝設有高溫石的表面直接烤焗，通常是硬質麵包，需要很高爐溫的麵包。如果焗盤沒有裝高溫石，可到相關店鋪訂造，把高溫石放入焗爐預熱至超熱，把麵糰放在上面烤焗；這樣，爐爆的效果最好。通常歐洲麵包或硬質麵包都需要用這方法去烤焗，又或者到五金店請店員切割一塊焗盤大小的厚鋁片，和高溫石的方法一樣 ，可以造歐洲包，當然這做法沒法和專業焗爐相比。

焗盤烘焙

就是麵糰造型後放在焗盤上作最後發酵，然後放入爐烤焗，通常如果麵糰包餡，容易滲出汁液或容易溶化材料例如芝士等，就需要用焗盤烤焗麵糰，這是最多用來烤焗麵包的方法。

模具烘焙

如果我們要造吐司（即方包），或是山形吐司（即枕頭包）或者造一些特別形狀的麵包時，我們會利用模具。市面有大大小小的模具，究竟我們要用多少麵糰才可入填滿呢？烘焙界會用比容積來計算，我們先來計算模具的體積：

方形體積是長度 x 寬度 x 高度；圓形體積是 3.1416 x 半徑 2 次方；不規則形狀可以用水去量度，得出的毫米就是體積，知道了體積後我們就可以計算了。我們會使用容積率來計算，容積率是麵包職人長久以來造入模麵包所得出的一些數據，一般吐司比容積率由 3.6-4.4，但可依照配方、特殊模具及個人喜好口感調整。容積率越小所使用麵糰越多，容積率越大所使用麵糰越小 。

現今的麵粉和材料都加入添加劑令麵糰改善，變得更有彈性，膨脹力更好，體積更大；所以所使用的麵糰就會越來越少，來減低成本，比容積率可大於 4，麵包就會很輕。

同一模具使用較小的比容積率，麵糰就越用得多，口感亦越紮實，一般購買吐司模時，會看到這些標示法：

以生麵糰的重量：

例如：1 斤、1.5 斤、12 兩、450 克……等。

其實依照這種重量標示算是比較直覺簡單的，由生麵糰重量即計算所需的麵粉重量。

只是各地的重量標示不同，要注意一下：

日本 1 斤，指的是英斤，也就是磅，1 磅約等於 450 克。

台灣 1 斤，指的是台斤，等於 16 兩，1 斤約等於 600 克。

在日本規定市售的吐司麵包 1 斤不可小於 340 克。

如果造山型吐司，麵糰重量＝體積 ÷4。

如果是一般角食（即頂角吐司）白吐司，麵糰重量＝體積 ÷3.8。

我較喜歡紮實口感的吐司，所以我會以 3.8 為白吐司的容積率。有人問過為何 450 克的吐司模要放 600 克的麵糰？就是比容積的關係了。

常見問題

Q 麵糰會斷筋嗎？

會的，當麵糰溫度超過 28℃時，如再搓揉或攪打，麵筋便會因過熱而變軟，搓揉或攪打令過軟的組織散掉，便斷筋了。斷筋的麵糰很油亮，沒有彈性，很黏手，再也不能救回了。

Q 為甚麼要用鮮酵母？

鮮酵母效力穩定，容易辨認是否壞掉，後發力好，氣味亦好，所以我比較喜歡使用。

Q 很多朋友查詢的問題：怎樣計算可於早上焗好新鮮麵包給家人吃？

我會説其實是可行的，但要凌晨 3 時起床預備到 7 時就有新鮮麵包吃囉。可是，都市人都要上班，睡眠時間是奢侈品，怎可能每天這樣；解決方法有的，就是投資一部可預設調節溫度的發酵箱，就可以將已造型的麵糰放進發酵箱，清早麵糰就發酵好，可以烤焗了，可是這可預設調節溫度的發酵箱動輒數萬元。沒可能買那麼專業的發酵箱時，我們都可以將已造型的麵糰放入雪櫃，冷凍麵糰，直至要吃麵包時拿麵糰出來解凍，再等待發酵，然後烤焗。但這方法由解凍至入爐需要 2-3 小時，也快不了許多。還是把麵包造好後放雪櫃儲存，吃時翻熱吧！

Q 造麵包中途要離開怎辦？

在發酵時、滚圓休息時或最後發酵時也可以將麵糰放雪櫃，暫時拖延一下時間，但也要儘快回來繼續處理。如果要離開時間頗長，那麼在滚圓休息或造型後把麵糰放冰格冷凍，直至回來再回溫繼續造。

Q 造麵包一定要用高筋麵粉嗎？

不一定的，有些朋友對小麥敏感的，可以使用無麩質麵粉或者思貝爾小麥粉製作麵包。亦可以用其他粉類，例如黑麥粉、全麥粉、粟米粉或米粉製作麵包。

Q 吐司為甚麼會收腰？

吐司焗不熟透、未有及時脱模或水份太高都會令吐司收腰。

圓形小餐包

麵糰	百分比（%）	重量（克）
高筋麵粉	100%	250 克
海鹽	2%	5 克
脫脂奶粉	3%	7.5 克
砂糖	10%	25 克
蛋	10%	25 克
鮮酵母	3%	7.5 克
水	55%	137.5 克
無鹽牛油	8%	20 克
	191%	**477.5 克**

準備：在搓揉開始前，請確保工作枱上是完全乾燥清潔，而且沒有雜物會妨礙到自己。預備一些手粉可以防止麵糰黏住枱面，手粉最好使用高筋麵粉，取其爽身和不黏手。

秤重：計算好所需材料份量，使用精確的儀器例如電子磅來量好各種材料。

混合：先將鮮酵母溶於食譜份量的部份水中，酵母溶解後好像陶泥水狀態，再把酵母水放入麵粉、餘下水、糖、蛋、鹽等等材料在大碗內以膠刮混合至不見到水份。如果用速效酵母，只要將酵母灑在麵粉內便可以搓揉，在搓揉中顆粒細小的速效酵母很快會溶解。（圖 1 - 3）

將濕麵糰放在工作枱面，開始用手搓揉。我們前後腳站立，用身體重量經雙手壓在麵糰進行搓揉。

搓揉：搓揉初期像洗衣服一樣按着麵糰前後推動，麵糰這時會很黏手，但請不要亂加麵粉，要繼續搓揉，動作要快，很快麵糰就會開始沒有那麼黏手，大約 6-8 分鐘，我們展開麵糰，有一點筋性此時可以投油了。（圖 4 - 9）

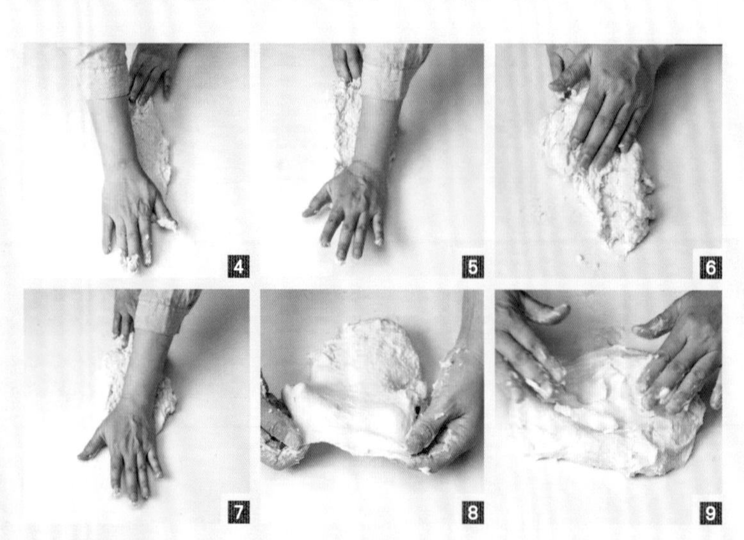

Video
搓揉：初期

投油：將麵糰打開，將室溫軟度無鹽牛油抹在麵糰上，把麵糰抓捏，讓牛油被麵糰吸收後便可以撻麵糰了。撻麵糰可以讓麵糰更快出筋膜，此時麵糰雖然已經吸收了水份和不算黏手，但我們如果用手心去接觸它時仍然會很黏，因為手心有汗水和熱力，所以在撻麵時只用指尖輕抓着麵糰的一小部份，然後摔出去枱面，再摺疊麵糰，直至麵糰光滑，有摺疊的動作，麵糰便更快起筋了。（圖 10 - 16）

Video
撻揉：投油後

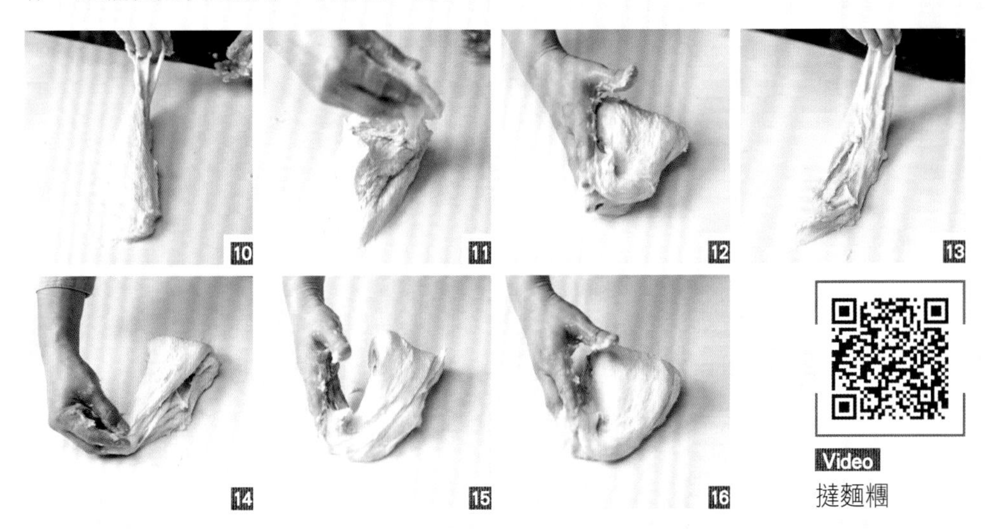

Video
撻麵糰

測試筋膜：將麵糰搓至光滑，有薄膜後可測試麵糰筋膜，切一小塊麵糰，用手指尖沾少許手粉，用手指輕輕拉開麵糰，觀察打開了的麵筋，如果看到麵糰筋膜的穿孔沒有鋸齒狀，洞邊圓滑，而用手彈時彈性足夠，就可以量度麵糰溫度，如果溫度未夠便要繼續搓揉至達到 24-26℃，如果已達到 24-26℃便完成了。（圖 17 - 19）

麵糰溫度超過 1-2 度，問題應該不大的，可以在發酵時調整時間，但麵糰超過 28℃時，最好立即將它稍為降溫，例如放它在冰涼的枱面或用冰墊讓它降回 24-26℃，否則繼續搓揉就很容易因過熱而斷筋。

整個搓揉大約在 15-20 分鐘內完成。

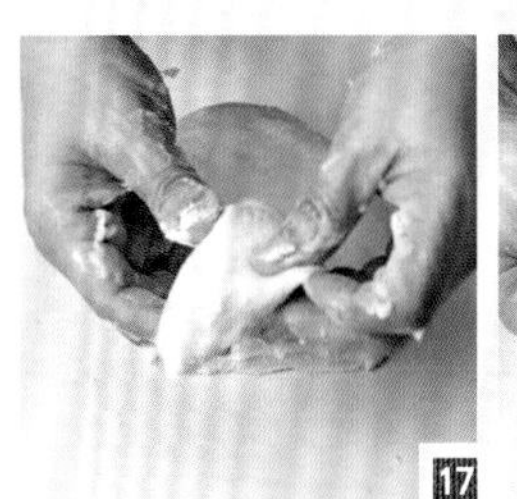

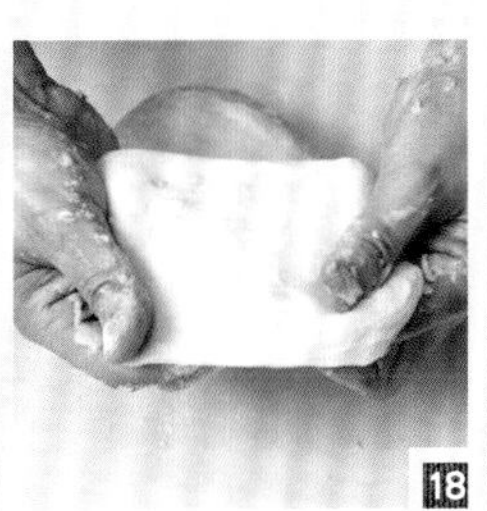

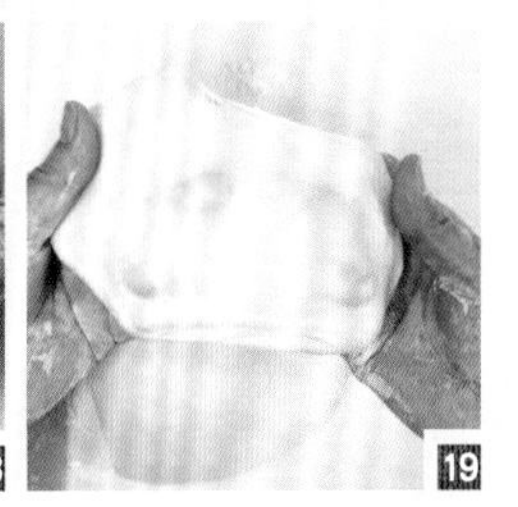

Video
麵糰筋膜測試

發酵：將搓好的麵糰滾成大圓球狀，收口捏好，這樣可以保存麵糰內的二氧化碳，放室溫（23-30℃）發酵約 45-60 分鐘，發酵溫度最好維持在 28-32℃；麵糰在室溫 45-60 分鐘後，會升溫 0.8-1℃，到達酵母最活躍的溫度，這時酵母就會分裂出很多倍了。（圖 20 - 25）

不時要觀察麵糰，如果 30 分鐘麵糰未有反應，可能你的酵母已死掉，麵糰就沒用了。

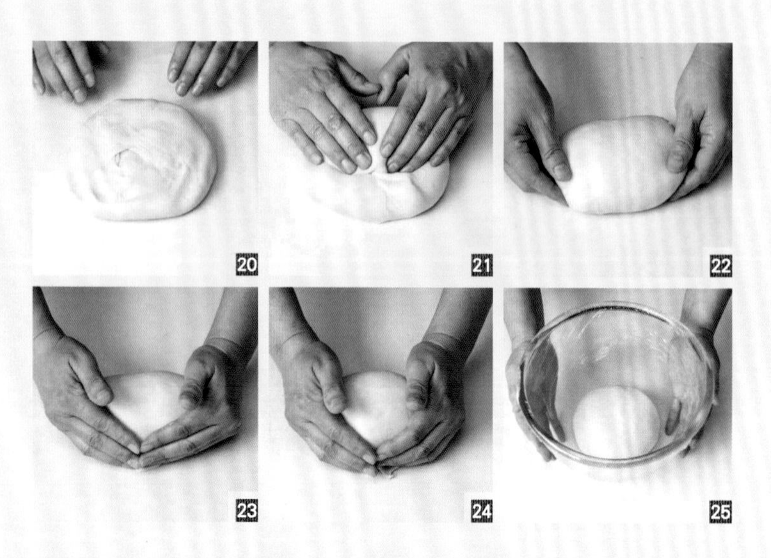

測試：大約 45 分鐘後可以測試麵糰，將手指沾上麵粉，插入麵糰中，慢慢戳至底部，戳孔沒有立刻縮小回彈，維持形狀即發酵完畢。（圖 26 - 27）

如果戳孔立即回彈即未發酵完成，要再多發酵 5-10 分鐘，重複測試麵糰直到發酵完成。（圖 28）

但如果戳孔立即穿掉、氣體洩出，即表示麵糰已過度發酵，不能再用來造麵包了。（圖 29）

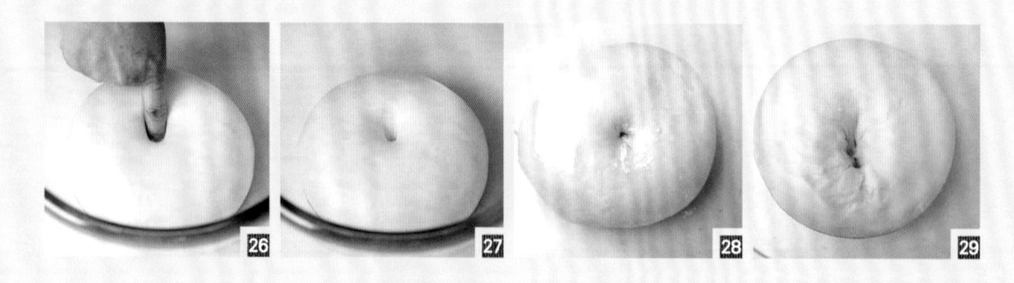

排氣、分割：第一次發酵完成，把發酵完畢的麵糰用膠刮刮出來，弄成方形，這時麵糰內的氣體會被排出，不用刻意去排氣；切成長條形再切細，這樣切法比較整齊和有效率。（圖 30）

Video
分割麵糰

利用電子磅秤重，將麵糰平均分切成等份，切成等份的小麵糰才容易控制整盤麵包的出爐時間。（圖 31）

在分切麵糰時要盡量快手，因為麵糰尚在發酵中，如果整份麵糰分切的時間太久，麵糰筋膜變得軟弱而容易被破壞，造不到好麵包。

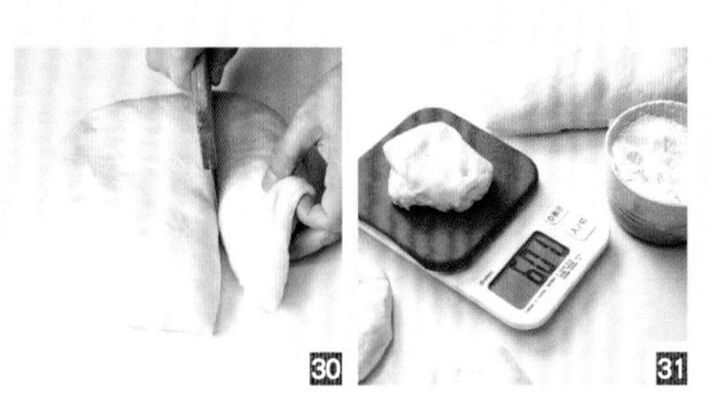

滾圓：視乎麵糰黏度在工作枱上撒少許手粉，把麵糰輕輕按扁，排出氣體，令內裏組織細緻、平均。

Video
滾圓麵糰

排出氣體後將麵糰光滑表面反過來向枱面，注意不要讓表面弄花，然後捲起小麵糰，整隻手的手指尖和手腕放鬆成爪形按在小麵糰上，貼緊枱面，前後移動幾下，這時小麵糰會成卷形被拉緊，然後繼續向左右移動幾下，再順時針轉幾下成鬆緊有致的滾圓。重複滾圓動作至所有麵糰滾好，放膠盒內，用蓋蓋着，以免麵糰表面被吹乾。（圖 32 - 38）滾好的麵糰應該鬆緊有致，底部收口緊密，如果收口不緊可用指尖捏緊。

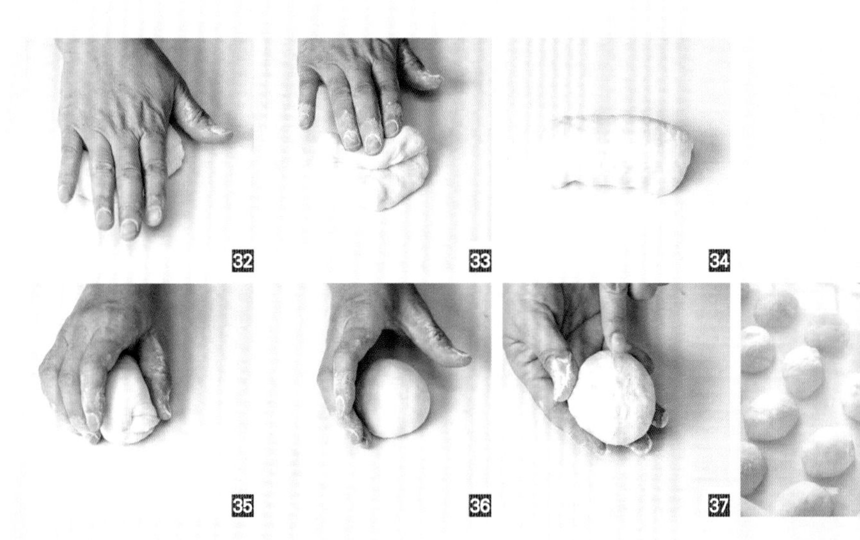

讓麵糰鬆弛約 15 分鐘，以方便造型；麵糰在滾圓後麵筋會變得緊縮，很難即時造型，讓麵糰鬆弛約 15-20 分鐘才容易造型。

要注意麵糰鬆弛時間，如果要離開一段小時間，最好將滾好麵糰放入雪櫃去拖延時間。

要注意手粉不能用過多，多餘的手粉會黏在麵糰裏外，讓滾圓時麵糰不能成糰，變得龜裂和有硬塊。

滾圓也可以在手心滾的啊！

滾圓是造圓形、欖形、三角形麵包的基本動作，一定要練熟啊！

Video
在手上滾圓

Video
滾圓透視

造圓形麵包的基本動作（圖 39 - 42）

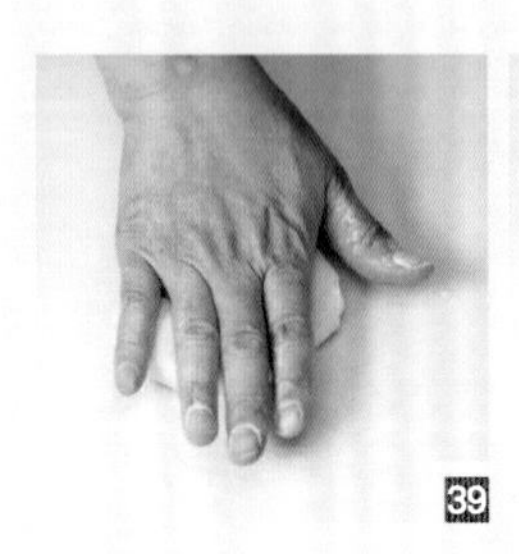
39

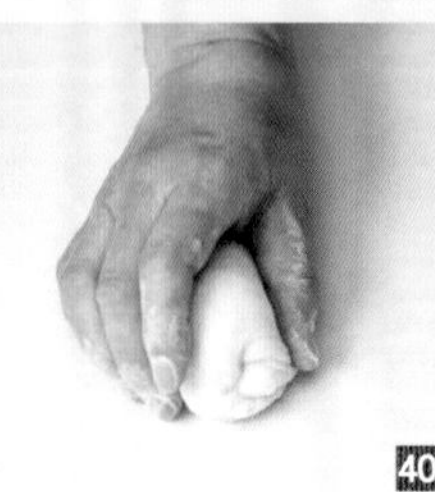
40

41

42

造欖形麵包的基本動作（圖 43 - 50）

Video
橄欖形麵包造型

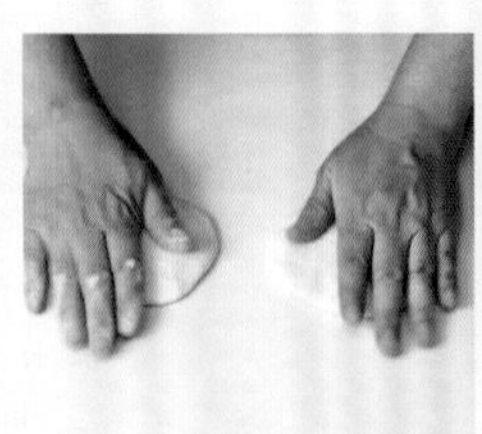
43

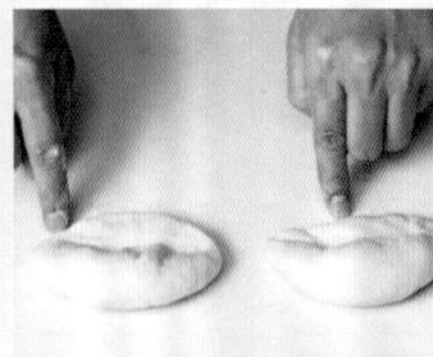
44

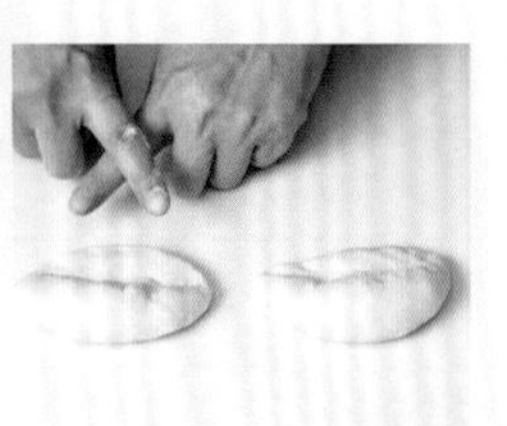
45

46

47

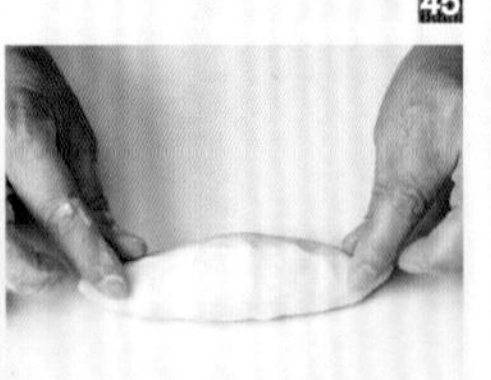
48

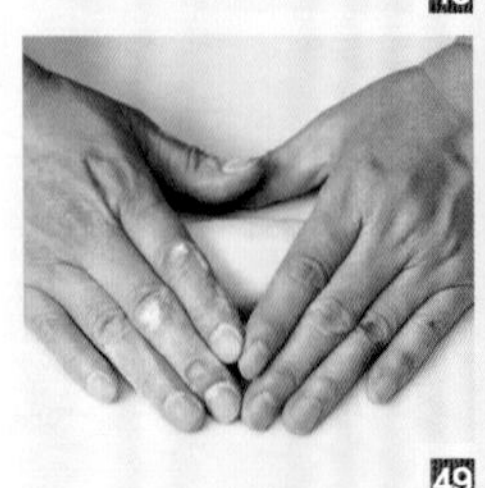
49

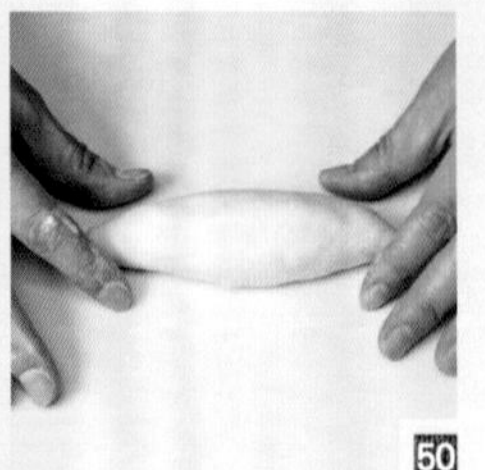
50

造三角形麵包的基本動作（圖 51 - 58）

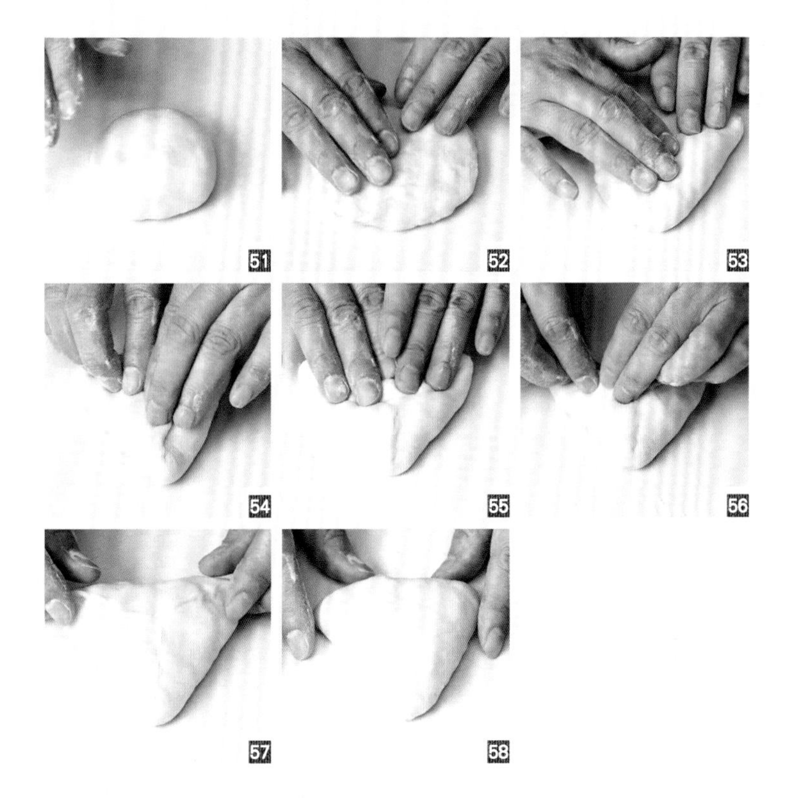

Video
三角形麵包
造型

除了滾圓，造長條形麵包、吐司等形狀就要用 Handsquare 這預備動作了，麵糰切好後捲摺成長形便可。（圖 59 - 61）

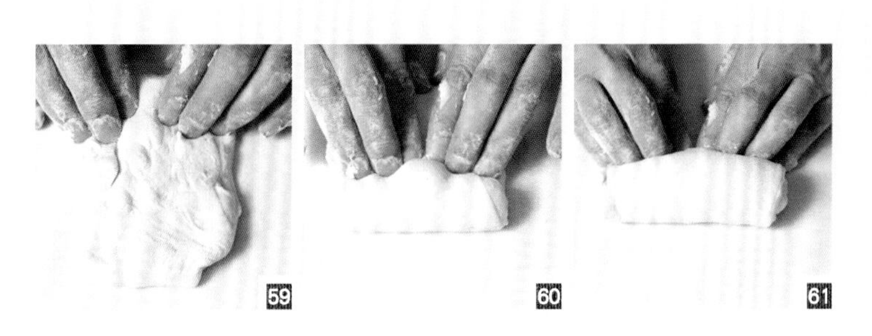

Video
Handsquare

造型：桌上撒少許手粉（用高筋麵粉），把麵糰輕輕按扁，排出氣體，重覆滾圓動作做成圓餐包，或摺出欖形、三角形餐包，放在焗盤上作最後發酵。

麵糰造型時要把大氣泡拍掉，小氣泡是沒問題的，滾圓時要力度鬆緊有度，太鬆麵包會形狀不美，有縐紋，不緊緻；力度太大麵糰會太緊，麵糰表面會被扯爛，麵包成品亦會變形。

注意每個小麵糰的距離，麵糰會發大，要有適當位置才不會黏在一起。

最後發酵：小麵糰要經最後發酵至適當體積才可入爐烤焗，因應麵糰溫度(24-26℃)，發酵溫度最好維持在 28-32℃，是酵母最活躍的溫度，發酵大約 30-45 分鐘，因應室溫而和實際發酵情況而定，要注意麵糰濕度，間中為麵糰噴水保濕；可用手指輕輕按麵糰測試，最後發酵完畢，麵糰發大約一倍。（圖 62）

在麵糰發酵時，可以預熱焗爐了。

62

裝飾：麵糰發酵好後可以入爐了，入爐前可以作點不同裝飾，例如上不同的光面，放不同的飾面，用不同模版上粉或剘花，完成裝飾。見後頁的上光、飾面、剘花、印模。

烤焗：一般烤焗的爐溫大約在 180-200℃，烤焗時間因應麵糰的大小而定。

- 40-50 克的麵糰約 10-12 分鐘
- 60-90 克的麵糰約 13-15 分鐘
- 100-150 克的麵糰約 16-18 分鐘
- 160-200 克的麵糰約 19-22 分鐘

如果焗至中途發現麵包上色不佳，可以轉動焗盤或將焗盤上下移動讓麵包上色得更好。

出爐：麵包焗好後，用手按下應該立即回彈，而且有香氣，這時可以出爐，否則多加 1-2 分鐘，不應焗得太久，令至麵包皮過硬和包肉乾涸。

Video
小餐包出爐

上光

麵包作最後發酵後，我們可以為它造點裝飾，令平凡的麵包變得不一樣；我們可以塗上不同液體替麵包上光，令它亮晶晶起來，之後再放上不同飾面，令它吃起來口感更豐富。不想上光的話可以剘上不同花紋當作自己的個性簽名。亦可以用不同版模在麵糰上篩上不同粉類，又是另類的型格裝飾。

上光，基本上以上蛋最普遍，其餘的可從圖中看到不同的光令度，味道只有糖水會有少許甜味，其餘是沒有味道的。

飾面

替麵糰上光後，我們還可以在上面加不同的飾面，讓平凡的麵包變得不一樣。

未烘焙加上飾面的麵糰（左），不同的飾面材料（右）

以下為大家介紹不同的裝飾品項

❶ 芝麻
❷ 芝士、香草
❸ 黑麥粉
❹ 葵花子
❺ 肉桂砂糖
❻ 罌粟子
❼ 杏仁片
❽ 芝士
❾ 燕麥片
❿ 粟米粉

1
2
3
4
5
6
7
8
9
10

剔花

不喜歡上光或飾面，可以用剪刀或剔刀在麵糰上來趟個性簽名，只要你喜歡的就可以劃上去了。

印模

你可以自己設計不同的印模，作為你專用的個性簽名。

Different ways of breadmaking

There are many ways to make bread. I'm just introducing a few popular methods here.

• Direct method

As its name suggests, you add all ingredients directly, knead or mix them into dough in a single step. The dough is then left to proof for a short period of time at room temperature before shaped and baked. This method has the advantages of short proofing time and easy simple workflow. It takes less time to make the bread and the flavours of the bread are showcased directly. On the down side, the dough contains less moisture so that the bread tends to over-ferment easily. The dough is lacking resilience and it doesn't have enough time to develop characteristic flavours and aromas. Some bakers remedy the last disadvantage by putting the direct dough in a fridge at 0 to 5°C for low-temperature proofing. After proofing in the fridge for 24 to 72 hours, the dough would have developed flavours and aromas.

Recipes in this book that use direct method: *Plain round rolls, Braided bread #3, 5 and 8, Cocoa buns with orange custard filling*

• Direct method – punching down

This is a variation of the direct method, but the dough needs more time to proof and to develop flavours and gluten structure. After the direct dough has been risen for a certain period of time, the rise will plateau out as its temperature rises and the alcohol and acetobacter have accumulated too much for the dough to keep expanding. At this point, it helps to flip the dough to make the dough more stretchable and to increase the number of yeast cells. Bread fermented this way tends to be more chewy and springy than that made with direct method.

Recipes in this book that uses direct method - punching down: *Wholemeal bread, Vegetable flat bread, Raisin buns, Chicken a la king buns and Sweet corn salad buns.*

• Sponge method

A sponge is a preferment made with part of the dry ingredients, water and yeast of the main dough ahead of time and allowed to proof and mature. The sponge is then added to the rest of the bread ingredients, mixed or kneaded into dough before risen and baked. There are many kinds of sponge recipes, such as 50% sponge, 70% sponge, 80% sponge, sugar sponge, and room temperature sponge etc.

Sponge method is sometimes called the two-phase method as the dough ingredients are mixed in two batches. The sponge is mixed first and allowed to rise for a period of time before the rest of the ingredients are added and kneaded into dough. Yeast activity is undermined by high sugar, salt and grease levels. Using a sponge can let the yeast grow properly and actively in the first phase and let the moisture permeate into the flour particles, so that the dough is moister. Then when other ingredients are added in the second phase, all the air bubbles will burst and the dough will be soft with fine crumbs.

The sponge needs to proof at proper temperature for the appropriate period of time. Under-fermented or over-fermented sponge would make the bread less tasty, whereas the fermentation temperature is crucial. A dough with sponge tends to rise more quickly, with bigger volume and softer texture. The dough won't over-ferment easily and is easier to handle. Among various sponge recipe, the 70% sponge is the most popular. It means the ratio of the flour in the sponge and that in the main dough is 7 to 3 by weight. Then salt, yeast and water are added to the sponge and left at room temperature or a low temperature for fermentation, before added to the main dough and kneaded.

Sugar sponge is made up of part of the sugar, flour, yeast and water from the main dough. It is allowed to proof before added to the rest of the ingredients. It is commonly used in sweet bread with high sugar content.

Recipe in this book that uses the sponge method: *Shredded coconut buns / loaves.*

• Old dough method, pâte fermentée / fermented dough

This method involves adding dough that has fermented once into fresh ingredients. The old dough has a long time to develop flavours and aromas, so that the bread will end up being full of natural wheat fragrance and being more chewy and soft, with finer crumbs. Researches show that yeast needs 45 minutes in the dough to adapt to the environment before releasing carbon dioxide and alcohol. Using old dough is not one way to work around the environments that hinder yeast activity (such as high sugar or salt content), but also a way to skip the adaptation time. When the old dough ferments the first time, the yeast have already passed that 45 minutes period and let the yeast grow properly and actively. This is the key difference between the direct method and the old dough method.

• Old dough

Old dough is an excellent all-natural additive that enhances the bread's texture and moisture content. The bread made with old dough tends to stay springy and moist for a longer period of time after baked. Old dough usually contains 15 to 30% of the flour in the whole dough. Contrary to sponge, old dough is a dough itself and it does not affect the baker's percentage of the main dough. Thus, you don't need to readjust the amounts listed on a recipe if you use old dough. However, a sponge would change the baker's percentage of other ingredients and you cannot readjust the amounts of ingredients haphazardly.

You may make the old dough first (even with a little sugar or oil) and divide it into 50 g pieces. Then freeze them. When you make bread, take the old dough out of the freezer and leave it to thaw at room temperature for 30 minutes before using. You may even wrap it in a plastic bag and put it in warm water to thaw. You may add old dough to almost any bread you make. If you're not making a bulk volume of bread, you don't need to alter the amount of ingredients at all. For those who knead by hands, just tear the old dough into small pieces and soak them in the volume of water specified by the recipe briefly until soft before kneading. They will incorporate more easily with the main dough that way.

Old dough	Baker's percentage(%)	Weight(g)
Bread flour	100%	250 g
Water	60%	150 g
Fresh yeast	0.4%	1g
Salt	0.4%	1g

Method:

Knead all ingredients together. Leave the dough to ferment at room temperature for at least 4 hours. Or, you may leave it to proof in the fridge for over 10 hours. Use the amount you need right away. Divide the leftover into equal portions of the weight you need for each baking session. Roll each portion round and wrap each up in cling film. Keep in the freezer and they last for 2 to 3 months. Thaw at room temperature before use.

• Pâte fermentée (French old dough)

Pâte fermentée and old dough are both fermented dough, making up 15 to 30% of the main dough by weight. It is essentially the same thing as old dough, but the flour used must be Type 55 French flour with additional malt extract is used. Pâte fermentée develops an aroma and flavours unique to European-style crusty bread. That being said, you may still use it to make dinner rolls. However, you cannot use regular old dough to replace pâte fermentée when making European-style crusty bread.

Pâte fermentée	Baker's percentage(%)	Weight(g)
French flour Type 55	100%	100 g
Sea salt	1%	1 g
Malt extract	0.5%	0.5 g
Fresh yeast	0.5%	0.5 g
Water	65%	65 g

The dough should be at 24°C after kneaded. Leave it to proof at room temperature for 90 minutes. Then transfer into a fridge at 0 to 5°C and leave it to proof for 6 to 8 hours. Alternatively, put the dough straight into a fridge at 0 to 5°C right after kneading and leave it there for 12 to 24 hours before use.

Recipes in this book that use old dough or pâte fermentée: *Flower-shaped buns with purple sweet potato filling, Cheese rolls / Tuna rolls.*

• Yudane (utane) method

"Yudane" means hot water starter in Japanese. The word "yu" means hot water or hot spring. This method is to mix flour with water into a paste and then heated to a specific temperature. Or, water is heated up to a certain temperature and added to flour into a paste. The heat causes the starch to gelatinize and this paste is known as Yudane. Then it is added to other dough ingredients and mixed or kneaded into dough. After gelatinization, the starch is capable to hold more water to make the bread moister. The air bubbles in the dough is also smaller in sizes so that the crumbs are finer in texture. The end-product is chewy, fluffy and moist. It also stay moist and fresh for a longer period of time. Popular Yudane recipes include 65°C Yudane and 85°C Yudane.

Recipes in this book that use Yudane method: *Sausage rolls, Ham, cheese and onion loaf*

• Poolish

It is a preferment with higher water content which was originated from Poland before being introduced to France. The basic recipe is 100% water mixed with 100% flour with a small amount of yeast, fermented at room temperature, or at 0 to 5°C for 12 to 24 hours. The yeast has more time to act on the starch and protein in the dough this way so that the dough is more stretchable. It takes less time to knead the dough because it turns smooth very quickly with poolish. The bread also tastes richer and has longer shelf life without drying out quickly. Poolish fermented at room temperature gives the bread a mild sweetness, whereas that fermented at low temperature develops richer flavours and more complex aromas.

Poolish has to ripe properly for the best results. Ripe poolish should have honeycomb structure and should be in milky white colour. It should be plump and round in shape, with a beautiful crack, exuding rich wheaty smell. Some people even use fruit juice or puree in poolish to give the bread fruity aromas and beautiful colours. However, fruits like papaya, pineapple or kiwi contain too much enzymes. Adding them to the poolish may hinder the fermentation process.

Recipe in this book that uses poolish: *Cranberry walnut loaves*

• Low-temperature retarded fermentation

This method is to keep 50% of the dough at 0 to 5°C to proof slowly in the fridge for 17 to 72 hours before mixed with other ingredients. Dough made this way takes less time to proof and the final proof tends to be powerful and substantial. The end-product has soft and moist crumbs with a hint of yeasty fragrance. The bread also has longer shelf life and stays soft and moist for a longer period of time.

For recipes, please refer to another title by Kin Chan Natural Breads Made Easy.

Plain Round Rolls

Ingredients	Baker's Percentage (%)	Weight (g)
bread flour	100%	250 g
sea salt	2%	5 g
skimmed milk powder	3%	7.5 g
sugar	10%	25 g
egg	10%	25 g
fresh yeast	3%	7.5 g
water	55%	137.5 g
unsalted butter	8%	20 g
	191%	**477.5 g**

Preparation

Before making and kneading the dough, make sure the countertop is clean and dry. Remove any items that would get in the way. Set aside some flour to sprinkle on the counter so that the dough doesn't stick. Preferably, use bread flour for this purpose because it's dry and doesn't stick to your hands as much.

Measurements

Weigh all the ingredients separately as listed, preferably with accurate devices such as a digital scale.

Mixing

Fresh yeast needs to be reactivated. Just dissolve the yeast in part of the water until the mixture resembles clay slurry. Then put the yeast mixture into a mixing bowl, and mix with flour, remaining water, sugar, egg, salt and all other ingredients (except butter). Use a rubber scraper to mix until no dry patch is visible. If you use instant yeast instead of fresh yeast, you don't need to reactivate it. Just add instant yeast directly to the rest of ingredients and start kneading. Instant yeast has smaller granules and they will dissolve quickly in the dough.
(pics. 1 - 3)

Transfer the wet dough onto a countertop and start kneading with your hands. Stand on the floor with one foot in front of the other. Knead by transferring your body weight onto the dough.

Kneading

To begin with, knead the dough as if you're washing clothes with a back and forth motion. The dough is very sticky at this stage and please resist the temptation to add more flour at this point. Just keep on kneading with swift and short motion. The dough won't feel as sticky very soon. After kneading for about 6 to 8 minutes, stretch the dough and it should have developed some gluten at this stage. Then add butter.
(pics. 4 - 9)

Adding butter

Press to flatten the dough. Spread unsalted butter that has been softened at room temperature on it. Squeeze and rub the dough until the butter is absorbed. Then slap and stretch the dough on the counter to quickly develop gluten. The dough should have absorbed all moisture and should not be sticky. But if you squeeze it with the centre of your palms, it still feels sticky because of the sweat and heat of your palms. Thus, you should just grab the dough with your fingertips when slapping it on the counter. Fold the dough and slap again repeatedly until the dough is smooth. Gluten develops quickly with this slapping and folding steps.
(pics. 10 - 16)

Windowpane test

When the dough is smooth, you can conduct the windowpane test to see if there is enough gluten in it. Cut a small piece of dough. Flour your fingertips. Stretch the dough with your fingers into a thin translucent membrane. Check the edges of the hole you make. If the edges are smooth not jagged, and if you feel the resilience when you tap on the membrane with your finger, you can measure the dough temperature. It is ready for the first proof in between 24 to 26°C. If it's under 24°C, keep on kneading. (pics. 17 - 19)

The dough is usually fine if it's one or two degrees above 26°C and you can make adjustment with the proofing time. However, if the dough is above 28°C, you should leave it on a cold counter or put it over an ice pack to bring the temperature down to between 24 to 26°C. Otherwise, the gluten structure tends to break easily if you keep on kneading at higher temperature.

The whole kneading step should last for 15 to 20 minutes.

First proofing

Roll the dough into a ball and tuck in the seam so that carbon dioxide gas can be trapped properly in the dough. Leave it at room temperature (23-30°C) to proof for 45 to 60 minutes. The temperature is best kept between 28 - 32°C. After rising for 45 to 60 minutes, the dough should be warmed up by 0.8 to 1°C. This is the temperature when the yeast is the most active and it divides and multiplies many times. (pics. 20 - 25)

Check on the dough from time to time. If the dough doesn't rise after 30 minutes, the yeast might be dead. The dough is no longer useable.

Ripe test

Test for ripeness after proofing for 45 minutes. Dip your finger into some flour. Poke the dough with it all the way to the bottom of the dough. If the indentation keeps its shape without shrinking or disappearing, it's ripe for punch down and shaping.(pics. 26 - 27)

If the indentation shrinks and disappears, the rising is not complete yet. Leave it for 5 to 10 more minutes and repeat the test again until it is ripe. (pics. 28)

If the dough shrinks drastically and deflates once your poke your finger in it, it has over-proofed. The dough is no longer useable. (pics. 29)

Punch down

After the first rise, remove the dough from the bowl with a rubber scraper. Shape it into a square. In the meantime, the gas bubbles in the dough are released so that you don't need to punch down by pressing the dough with your fist. Cut the dough into long strips and then into small pieces. The dough will be neater this way and you can divide the dough more efficiently. (pics. 30)

Dividing

Make sure each small piece of dough weighs the same with a digital scale.(pics. 31) It's important to keep them the same weight for uniform baking time in the oven at last. When you cut the dough, do it quickly. As the dough is still proofing while you perform all these steps, the gluten structure can be weakened if you take too much time to cut them.

Rolling round

Sprinkle some flour on the countertop if needed. Gently press the small piece of dough to release gas bubbles. That would give the bread a finer grain and even texture.

After releasing the bubbles, flip the dough upside-down so that the smooth side faces the counter. Keep the surface smooth and try not to make any marks on it. Roll up the dough by curling and fingers towards your wrists. Then roll the dough on the counter back and forth for a few times to tighten the dough. Then roll it to the left and right a few times. Finally, roll it clockwise for a few times into a tight ball. Repeat this step with all the cut dough pieces. Transfer them into a plastic box and cover the lid so that the dough balls won't dry up. (pics. 32 - 38) After rolled round, the dough should not too tight or too loose. The seam should be facing down and secured. If the seam is coming apart, pinch along it to secure again.

Leave them to rest for 15 minutes before shaping. The gluten will be tightened after rolled round. The dough will spring back and cannot be shaped easily. Thus, you should leave it for 15 to 20 minutes before shaping.

Pay attention to the resting time. If you have to leave the kitchen for a short while, make sure you keep the dough balls in a fridge to slow down the fermentation.

Do not use too much flour when rolling round the dough. Too much dry flour will stay on the surface of the dough and it prevents the formation of a uniform gluten structure. The dough will end up cracking with hard lumps here and there.

You may use the centre of your palms to roll the dough round.

Rolling round is the basis of making round (pics. 39 - 42), oval (pics. 43 - 50) or triangular buns (pics. 51 - 58). You should familiar yourself with this skill.

Besides rolling round, you'd need to handsquare the dough for baguettes or loaves. Just fold the dough into rectangular pieces after cutting them into uniform weights. (pics. 59 - 61)

Shaping

Sprinkle some bread flour onto the countertop. Gently press the dough balls to release gas bubbles. Roll them round again into round bread rolls. Or you may fold them into oval or triangular shapes. Arrange neatly on a baking tray for final proofing.

When you shape the dough, you should press it to release all the large bubbles. It's fine to keep the small bubbles. When you roll it round, you should pay attention to the force you use. If you do it too loosely, the bread won't keep it shape well and it tends to wrinkle after baked. If you do it too tightly, the gluten structure would be torn on the surface and it doesn't look smooth. The bread may also distort after baked.

Make sure you leave enough space between the dough when you put it on the tray. The dough will proof once more and will expand further. Make sure the rolls won't stick together at last.

Final proofing

The dough pieces need to proof once more to double their sizes before baked in an oven. If the dough is about 24 to 26°C in temperature, the best proofing temperature would be 28 to 32°C. It takes 30 to 45 minutes, depending on the temperature and the proofing situation. You should also make sure the dough doesn't dry out in the process. Spray water on it from time to time. Check for dampness by touching it with your finger. The dough is ready to be baked once it doubles its original size. (pics. 62)

While the dough is proofing, preheat the oven.

Decoration

You may decorate the dough before baking it – brush on egg wash for a glossy finish, garnish with seeds or oatmeal on top, sprinkle on flour with a stencil, or slash it with a bread lame. It's up to you.

Baking

The oven should be preheated to 180 to 200°C. The baking time depends on the weight of the dough.

- For dough pieces about 40 to 50 g each, bake for 10 to 12 minutes.
- For dough pieces about 60 to 90 g each, bake for 13 to 15 minutes.
- For dough pieces about 100 to 150 g each, bake for 16 to 18 minutes.
- For dough pieces about 160 to 200 g each, bake for 19 to 22 minutes.
- If you find the bread not browning evenly halfway through the baking time, you can turn the baking tray around or move it up or down in the oven.

Removing from oven

When is bread is properly baked, it should rebound after you press it with your hand. It should smell fragrant and it's ready to be served. If it's not ready yet, bake for 1 or 2 minutes more and check again. Do not overbake the bread. Otherwise, it will be too crusty and the crumbs will be too dry.

Glazing

You may brush on different liquids to add a sheen to the bread.

❶ dusted with flour
❷ brushed with whole egg
❸ brushed with egg white
❹ brushed with melted butter
❺ brushed with cream
❻ unglazed plain bread
❼ brushed with olive oil
❽ brushed with sweet glaze
❾ brushed with egg yolk

Toppings

You can put the different toppings on the dough to add textures.

⑩ sesame seeds
⑪ cheese and herb
⑫ rye flour
⑬ sunflower seed
⑭ cinnamon sugar
⑮ poppy seeds
⑯ almond flakes
⑰ cheese
⑱ oats
⑲ cornmeal

Scoring

If you don't want to add shine to the bread and don't want to garnish with other ingredients, you can simply score the bread with scissors or a bread lame as your signature.

Stencils

You can design different stencils and use them as your signature.

軟小餐包

直接法：學習不同造型
編織包 3、5、8 號花

麵糰材料	百分比（%）	重量（克）
高筋麵粉	100%	250 克
砂糖	10%	25 克
海鹽	2%	5 克
脫脂奶粉	4%	10 克
蛋黃	10%	25 克
鮮酵母	3%	7.5 克
水	55%	137.5 克
無鹽牛油	15%	37.5 克
	199%	**497.5 克**

麵糰做法：參考第 38 頁「圓形小餐包」的混合、搓揉、投油、測試筋膜、發酵、測試、排氣、分割和滾圓的步驟和圖 1-38。

Video
3 號花

造型：3 號花：桌上撒少許手粉，把麵糰輕輕按扁，排出氣體，再用擀麵棍擀薄成長舌形，上下摺疊起來成長條形，捏好收口，輕搓長成長辮子，將辮子打結，一邊預留稍長的尾部，頭尾交接成 3 號花。（圖 1 - 15）

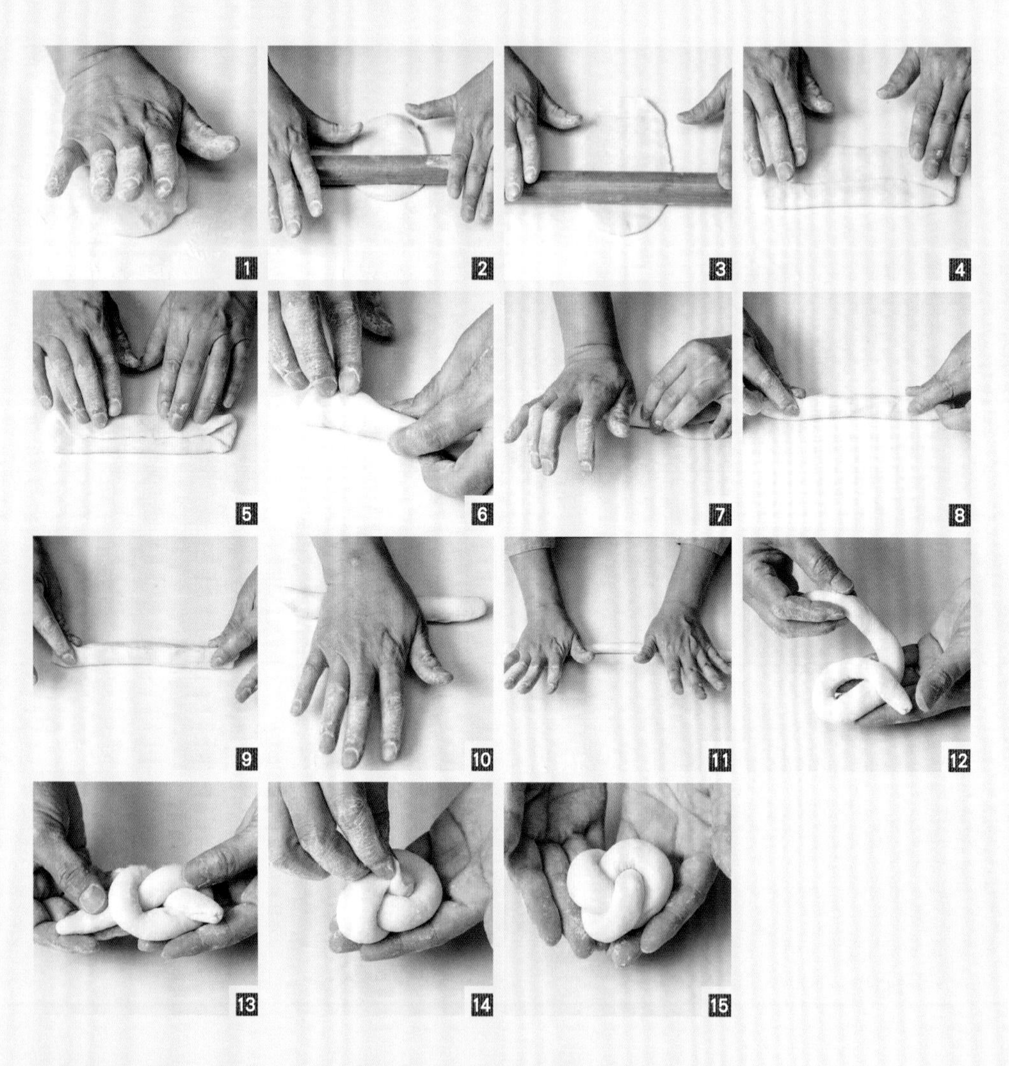

5 號花：將辮子打結，一邊預留比 3 號花長兩陪的尾部，在結圈穿過兩次，頭尾交接成 5 號花，在穿過結圈時，平均分佈好圈與圈的距離便會整齊美觀。（圖 16 - 21）

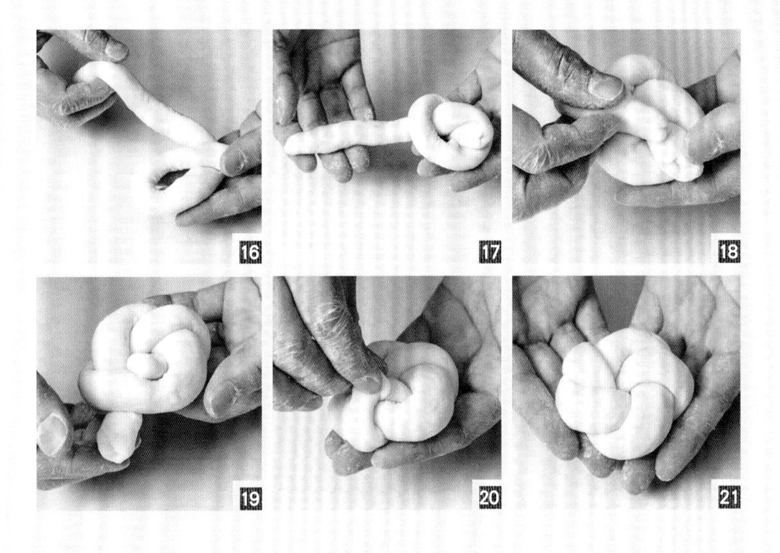
16 17 18 19 20 21

Video
5 號花

8 號花：請看相片示範。（圖 22 - 27）

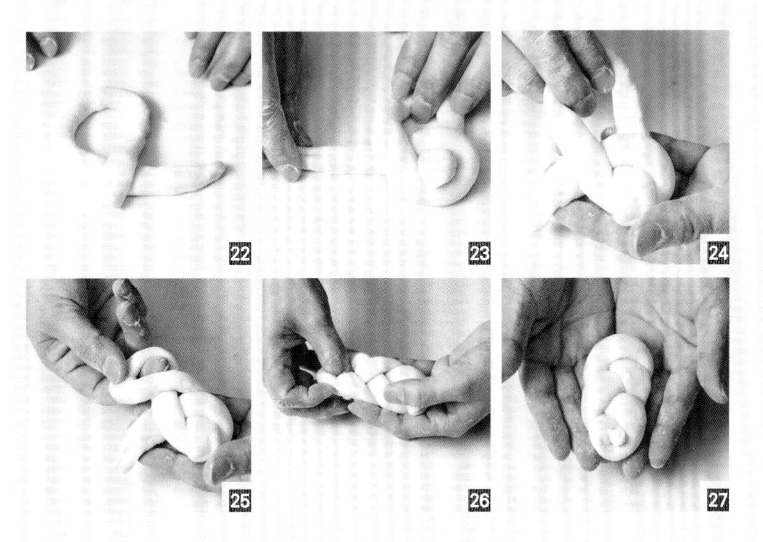
22 23 24 25 26 27

Video
8 號花

辮子一次過未必能搓長到所需長度，先搓第二、三條，回頭再繼續搓就可以了，但要留意不要把麵糰壓扁、壓斷，要保持麵糰圓渾；另外也要注意每次只摺 1 至 3 條辮子，因為麵糰很快會發酵膨脹，要再排氣使它回復光滑，很費時間。造完 1 至 3 條再繼續其他的，儘快造完。打結時不要拉扯麵糰，在辮子上灑少許手粉，不要令辮子黏着，破壞造型。

最後發酵：大約 20-25 分鐘，塗蛋液。

有紋理的造型麵糰最後發酵都不要太足，以免造型因過度膨脹而走樣。

烤焗：放入已預熱約 200-210℃焗爐，焗約 13-14 分鐘。

Braided Bread # 3, 5 and 8

Dough ingredients	Baker's Percentage (%)	Weight (g)
bread flour	100%	250 g
sugar	10%	25 g
sea salt	2%	5 g
skimmed milk powder	4%	10 g
egg yolk	10%	25 g
fresh yeast	3%	7.5 g
water	55%	137.5 g
unsalted butter	15%	37.5 g
	199%	**497.5 g**

Make the dough

Follow the steps in the recipe "Plain Round Rolls" on p.59 for mixing, kneading, adding butter, windowpane test, first proofing, ripe test, punch down, dividing and rolling round.

Shaping

Braided bread #3: Sprinkle bread flour on the counter. Gently press the dough to release air bubbles. Roll it into a long oval shape with a rolling pin. Fold along the short side to make a long log. Pinch to seal the seam. Roll on the counter into a long strip. Make a knot with the dough strip so that one loose end is longer than the other. Connect the loose ends at last. (pics. 1 - 15)

Braided bread #5: Roll the dough into a long strip. Make a knot with the strip so that one loose end is twice as long as you would with braided bread #3. Bring the long end through the knot twice and then connect the two loose ends together. When you bring the loose end through the knot, make sure you distribute the knots evenly so that the strands look like a five-petal flower. (pics. 16 - 21)

Braided bread #8: Please refer to the photos for method. (pics. 22 - 27)

You might not have kneaded each strand of dough to the perfect length at first. Thus, you should just roll out the second and the third strands first. Then go back to roll out the first one. Make sure you do not flatten the dough strand because they should be roughly round in cross section. When you roll them out, only do it up to three strands at one time. It's because they tend to proof very quickly. If you roll out too many strands at one time, the first one may have risen too much before you finish the last one and you have to punch down the first one and shape it again. It's such a waste of time. Thus, roll out three strands before rolling out others and try to do it as quickly as you can. When you braid the strands, do not pull and stretch them. Sprinkle some bread flour on the braided strands so that they won't stick to each other ruining the braids.

Final proofing

Let it proof in the tray for 20 to 25 minutes. Brush on egg wash.

Bread that is shaped to accentuate lines and forms should not be risen too fully in the final proofing. Otherwise, the lines and forms may distort if the dough happens to over-proofed.

Baking

Bake in a preheated oven at 200 to 210°C for 13 to 14 minutes.

全麥包

直接法：學習翻面及添加不同粉類

採用 30% 全麥粉

麵糰材料	百分比（%）	重量（克）
高筋麵粉	70%	224 克
全麥粉	30%	96 克
海鹽	2%	6.4 克
脫脂奶粉	3%	9.6 克
砂糖	8%	25.6 克
鮮酵母	3%	9.6 克
水	50%	160 克
無鹽牛油	6%	19.2 克
浸粉用水	24%	76.8 克
	196%	**627.2 克**

採用 50% 全麥粉

麵糰材料	百分比（%）	重量（克）
高筋麵粉	50%	160 克
全麥粉	50%	160 克
海鹽	2%	6.4 克
脫脂奶粉	3%	9.6 克
砂糖	8%	25.6 克
鮮酵母	3%	9.6 克
水	26%	83.2 克
無鹽牛油	6%	19.2 克
浸粉用水	44%	140.8 克
	192%	**614.4 克**

預備：用浸粉用水浸泡全麥粉，攪拌均勻放於 0-5℃約 8-12 小時。(圖 1 - 3)

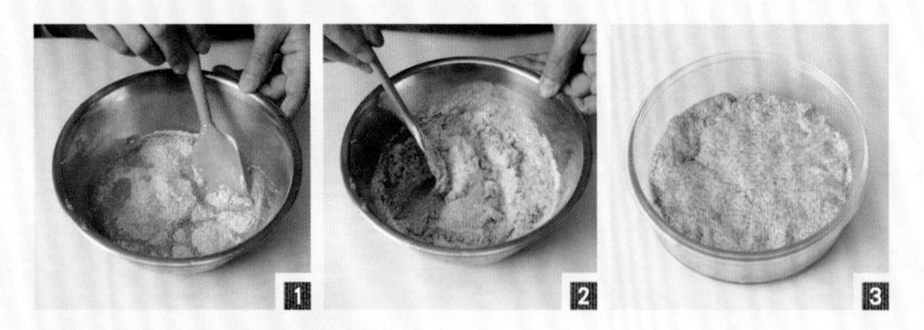

用 30% 全麥粉麵包

麵糰做法：參考第 38 頁「圓形小餐包」的混合、搓揉、投油、測試筋膜和發酵的步驟和圖 1-25。

發酵：麵糰發酵 60 分鐘後，將麵糰拿出來放枱面上翻面後再發酵 30 分鐘。

排氣：排氣，分割成等份麵糰、滾圓後（按第 41 頁「圓形小餐包」的排氣、分割、滾圓的做法），放雪櫃休息 15-20 分鐘至麵糰大約 15℃。

用 50% 全麥粉麵包

麵糰做法：參考第 38 頁「圓形小餐包」的混合、搓揉、投油、測試筋膜和發酵的步驟和圖 1-25。(將麵糰滾成大圓球。圖 4)

麵糰發酵 60 分鐘後，將麵糰拿出來放枱面上翻面後再發酵 30 分鐘。(圖 5 - 11)

Video
翻面

排氣：排氣後，分割成大約 200 克麵糰 6 個，Handsquare 後放膠盒休息 15-20 分鐘。（圖 12 - 14）

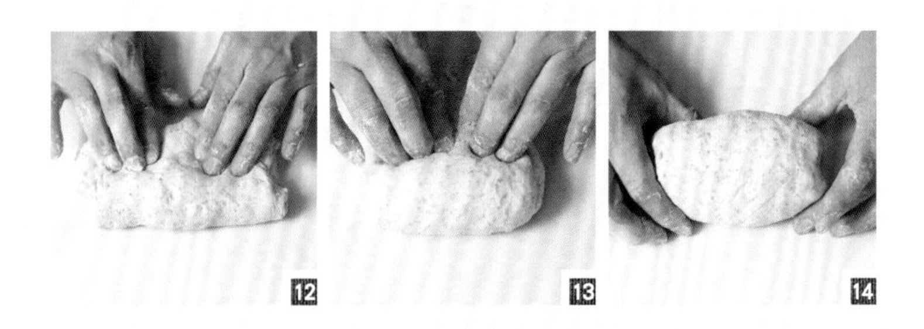
12 13 14

吐司模和蓋塗油。（圖 15）

15

把麵糰排氣，再用擀麵棍擀薄成長條，對摺，再擀薄，捲起放吐司盒內。（圖 16 - 29）

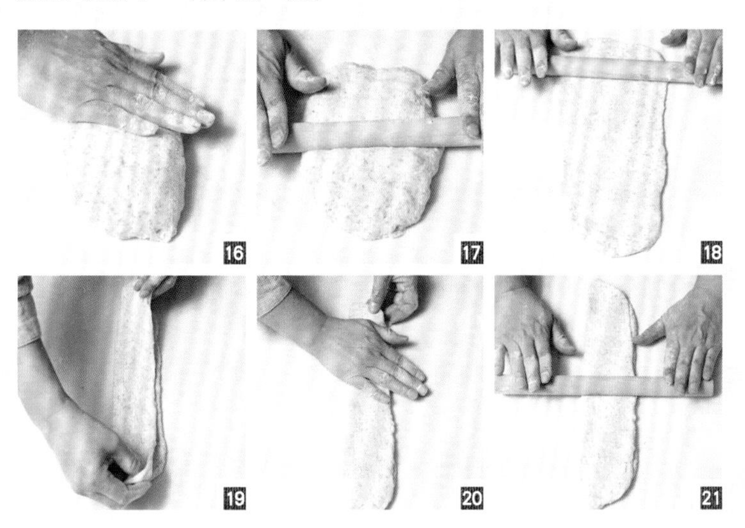
16 17 18 19 20 21

Video
全麥吐司造型

Video
全麥吐司入模

最後發酵：大約 45 分鐘，蓋上蓋。

烤焗：放入已預熱約 180℃焗爐，焗約 35-40 分鐘。

- 帶蓋吐司在出爐後未打開蓋前要先摔一下，讓熱空氣排出，讓吐司定型。
- 麵包中心溫度達 96℃才熟透。
- 左是用 30% 全麥粉吐司，右是用 50% 全麥粉吐司。（圖 30）

如果使用不同大小的模具，請先計算比容積：

- ▶ 7.5 厘米方形模大約用 120 克麵糰
- ▶ 17.7 × 6 × 8 厘米吐司模大約用 240 克麵糰
- ▶ 19.7 × 10.6 ×11 厘米吐司模大約用 600 克麵糰

Wholemeal Loaf

Direct Method: Punching down. Learn about adding dry ingredients.

30% wholemeal flour

Main dough	Baker's Percentage (%)	Weight (g)
bread flour	70%	224 g
wholemeal flour	30%	96 g
sea salt	2%	6.4 g
skimmed milk powder	3%	9.6 g
sugar	8%	25.6 g
fresh yeast	3%	9.6 g
water	50%	160 g
unsalted butter	6%	19.2 g
water (for soaking wholemeal flour)	24%	76.8 g
	196%	**627.2 g**

50% wholemeal flour

Main dough	Baker's Percentage (%)	Weight (g)
bread flour	50%	160 g
wholemeal flour	50%	160 g
sea salt	2%	6.4 g
skimmed milk powder	3%	9.6 g
sugar	8%	25.6 g
fresh yeast	3%	9.6 g
water	26%	83.2 g
unsalted butter	6%	19.2 g
water (for soaking wholemeal flour)	44%	140.8 g
	192%	**614.4 g**

Preparation

Soak the wholemeal flour in the water (for soaking wholemeal flour). Mix well and leave the mixture at 0 to 5°C for 8 to 12 hours. (pics. 1 - 3)

For 30% wholemeal loaf

Making the dough

Follow the steps of the recipe "Plain Round Rolls" for mixing, kneading, adding butter, windowpane test and first proofing.

Second proofing

After proofing for 60 minutes, punching down the dough on a countertop. Leave it to proof for more 30 minutes.

Punch down

Press the dough to release air bubbles. Divide the dough pieces about 60 g each and rolling round (follow the steps of the recipe "Plain Round Rolls" for punch down, dividing and rolling round). Leave them to rest in a fridge for 15 to 20 minutes until the dough cools down to 15°C.

For 50% wholemeal loaf

Making the dough

Follow the steps in the recipe "Plain Round Rolls" for mixing, kneading, adding butter, windowpane test and first proofing. (pics. 4)

After proofing for 60 minutes, punching down the dough on the countertop. Leave it to rise for 30 more minutes. (pics. 5 - 11)

Punch down

Press the dough to release air bubbles. Divide into 6 dough pieces weighing 200 g each. Handsquare them and put them in a plastic box. Cover the lid. Leave them to rest in a fridge for 15 to 20 minutes. (pics. 12 - 14)

Grease the loaf tin and the lid. (pics. 15)

Punch down the dough pieces again. Roll each out into a rectangle. Fold in half and roll it flat once more. Roll it up and press them into the loaf tin. (pics. 16 - 29)

Final proofing

Leave the dough to proof for 45 more minutes. Cover the lid.

Baking

Bake in a preheated oven at 180°C for 35 to 40 minutes.

- Remove the loaf from the oven together with the loaf tin without opening the lid. Tap the loaf tin once to release the hot air and to secure the shape of the loaf.
- The centre of the loaf should reach 96°C to be cooked through.
- 30% wholemeal loaf on the left; 50% wholemeal loaf on the right. (pics. 30)

If you use a loaf tin in different size, make sure you convert the amounts of ingredients according to the volume of the tin.

- For a 7.5 cm cube tin, you need 120 g of dough.
- For a loaf tin measuring 17.7 x 6 x 8 cm, you need 240 g of dough.
- For a loaf tin measuring 19.7 x 10.6 x 11 cm, you'd need 600 g of dough.

火腿芝士洋葱吐司

湯種法：吐司的不同變化

吐司是一種可塑性很高的麵包，如果只可以選一種麵包，我會選吐司，而我每天的早餐，也幾乎都吃吐司。

除了白吐司、全麥吐司，有餡料的吐司也很受歡迎。

湯種材料	百分比（%）	重量（克）
高筋麵粉	20 %	60 克
水（90℃）	30%	90 克
海鹽	0.5%	1.5 克

麵糰材料		
高筋麵粉	80%	240 克
砂糖	10%	30 克
海鹽	1.3%	3.9 克
脫脂奶粉	3%	9 克
蛋	10%	30 克
鮮酵母	3%	9 克
水	37%	111 克
無鹽牛油	8%	24 克
	202.8%	**608.4 克**

火腿	6 片
芝士片	6 片
沙律醬	適量
洋葱絲	半個份量

湯種製法：將水煮熱至 90℃，倒入筋粉中，攪拌至成糰，包着放雪櫃過夜，約 8-12 小時。（圖 1 - 3）

麵糰做法：參考第 38 頁「圓形小餐包」的混合、搓揉、投油、測試筋膜和發酵的步驟和圖 1-25。

發酵：麵糰發酵 60 分鐘後，將麵糰拿出來放枱面上翻面後再發酵 30 分鐘。（翻面做法看第 74 頁「全麥包」的圖 5-11）。

排氣：排氣，分割成 180 克麵糰，Handsquare，放雪櫃 15-20 分鐘至麵糰大約 15℃。（Handsquare 做法看第 43 頁「圓形小餐包」的圖 59-61）。

造型：排氣，擀長，放上沙律醬、洋葱絲、芝士和火腿片，捲起，放入塗油模內。（圖 4 - 13）

最後發酵：大約 40-45 分鐘，蓋上蓋子。

烤焗：放入已預熱約 180-200℃焗爐，焗約 35-40 分鐘。

Video

火腿芝士洋葱吐司入模

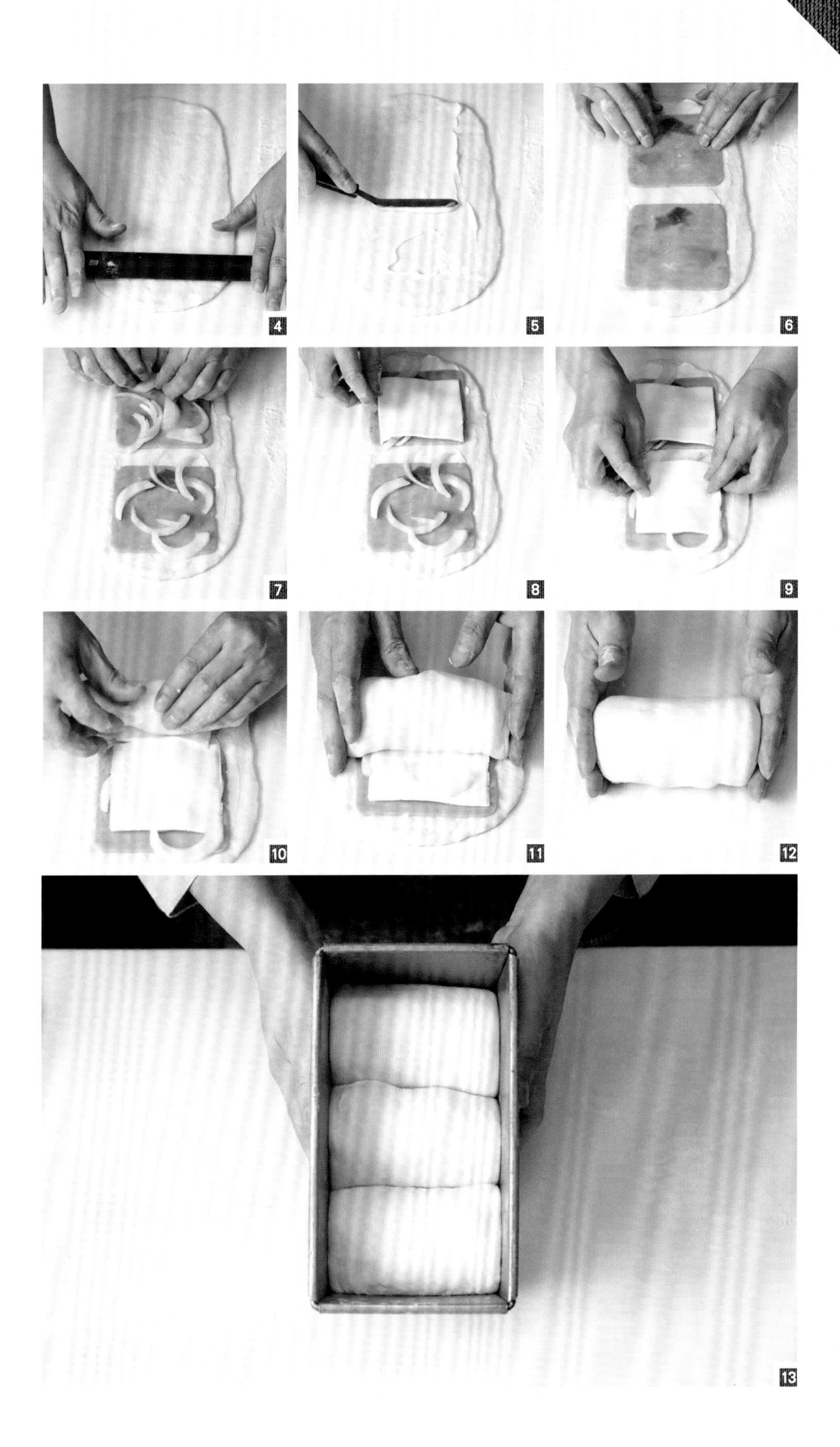
4
5
6
7
8
9
10
11
12
13

Ham, Cheese and Onion Loaf

A variation on classic sandwich loaves

Sandwich loaf is a versatile bread. If I can only choose one kind of bread for the rest of my life, I would pick sandwich loaf without a doubt. I have toast for breakfast almost every morning.

Apart from plain white loaf and wholemeal loaf, sandwich loaf with filling is also very popular.

Yudane (Utane)	Baker's Percentage (%)	Weight (g)
bread flour	20 %	60 g
water (at 90°C)	30%	90 g
sea salt	0.5%	1.5 g

Main dough		
bread flour	80%	240 g
sugar	10%	30 g
sea salt	1.3%	3.9 g
skimmed milk powder	3%	9 g
egg	10%	30 g
fresh yeast	3%	9 g
water	37%	111 g
unsalted butter	8%	24 g
	202.8%	**608.4 g**

6 slices cooked ham

6 slices cheddar cheese

creamy salad dressing

1/2 onion (shredded)

Making the Yudane

Boil water up to 90°C. Turn off the heat. Add bread flour and mix well. Knead until smooth and lump free. Wrap in cling film and refrigerate overnight for about 8 to 12 hours. (pics. 1 - 3)

Making the dough

Follow the steps of the recipe "Plain Round Rolls" for mixing, kneading, adding butter, windowpane test and first proofing.

Second Proofing

After proofing for 60 minutes. Punching down the dough on a countertop. Leave it for 30 minutes for second proofing. (Refer to the photo 5-11 in the recipe of "Wholemeal Loaf".)

Punch down

Press the dough to release air bubbles. Divide into dough pieces about 180 g each. Handsquare them. Leave them to rest in a fridge for 15 to 20 minutes until the dough cools down to 15°C. (Refer to the photo 59-61 in the recipe "Plain Round Rolls" for the method of handsquaring.)

Shaping

Punch down again. Roll out the dough into a long rectangle. Put salad dressing, onion, cheese and ham on top. Roll it up and press into a greased loaf tin.
(pics. 4 - 13)

Final proofing

Leave it for 40 to 45 minutes with the lid covered.

Baking

Bake in a preheated oven at 180 to 200°C for 35 to 40 minutes.

椰茸包/小吐司

中種法

這次示範的椰茸包有兩個造型，分別將麵糰切成 2 厘米厚放在紙兜內，焗後有漂亮圈紋的椰茸包；另一個是將麵糰捲起後，在麵糰中央切成兩份，再放進 18 厘米的牛油蛋糕模內，焗後成椰茸小吐司。

中種材料	百分比（%）	重量（克）
高筋麵粉	70%	210 克
奶	43%	129 克
鮮酵母	1%	3 克
砂糖	2%	6 克

麵糰材料		
高筋麵粉	30%	90 克
脫脂奶粉	6%	18 克
海鹽	1.5%	4.5 克
鮮酵母	2%	6 克
奶	6%	18 克
淡忌廉	6%	18 克
蛋	14%	42 克
無鹽牛油	10%	30 克
	191.5%	**574.5 克**

預備中種：將中種材料搓至光滑，用保鮮袋袋着，放 0-5℃雪櫃內約 8 小時。

混合：將麵糰材料和中種材料（除了無鹽牛油）放在大碗內，以膠刮混合至不見到水份（參考第 38 頁「圓形小餐包」的混合步驟圖 1-3）。

搓揉：搓揉至麵糰起筋可以放入無鹽牛油，按「圓形小餐包」的搓揉步驟圖 4-9，繼續搓麵糰至有薄膜，量度麵糰溫度。

發酵：將搓好麵糰滾成大圓球，收口捏好，放室溫（23-30℃）發酵約 30 分鐘。

排氣：排氣，分割成 140 克麵糰 3 個，Handsquare。

其餘麵糰 Handsquare，放雪櫃休息 15-20 分鐘至麵糰大約 15℃。

（Handsquare 做法看第 43 頁「圓形小餐包」的圖 59-61 ）。

椰茸包

將麵糰擀薄成長方形約 30 厘米 ×10 厘米，塗上餡料，捲起約 8 厘米直徑，收緊邊位，用刀切成大約 2 厘米厚，平放紙兜內。（圖 1 - 7）

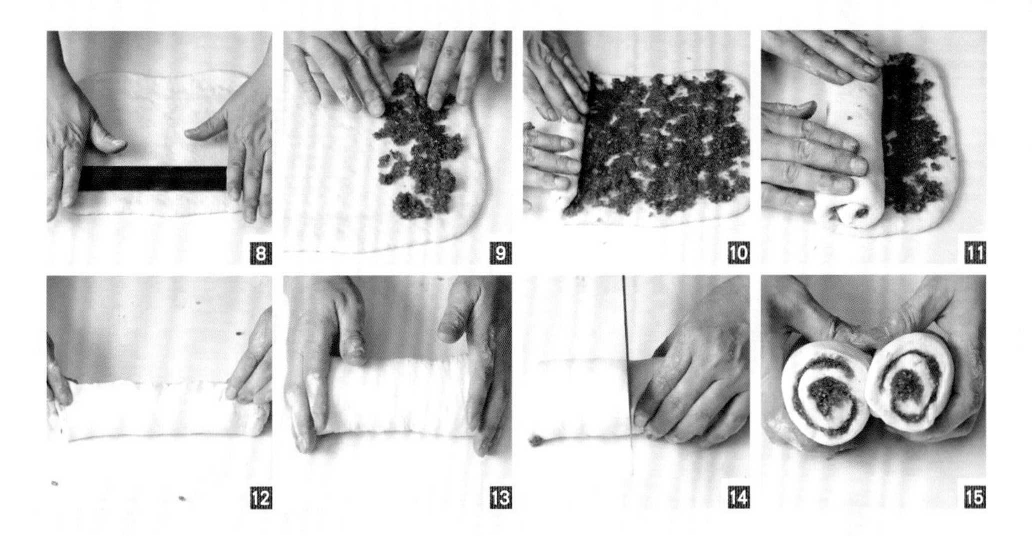

椰茸吐司

將 140 克麵糰擀薄成長方形約 30 厘米 ×10 厘米，塗上餡料，捲起，在中央切成兩份，放進兩個 18 厘米牛油蛋糕模內。（圖 8 - 15）

Video
椰蓉吐司

最後發酵：大約 20-25 分鐘，塗上蛋液。

烤焗：椰茸包放入已預熱約 180℃焗爐，焗約 15 分鐘；椰茸吐司放入已預熱約 180℃焗爐，焗約 24 分鐘。

椰茸餡做法

材料	重量(克)
椰茸	150
砂糖	125
蛋	70
溶牛油	50
* 即溶咖啡粉	8

* 不喜歡咖啡味可免去

做法：
將所有材料混合（圖 16 - 17）

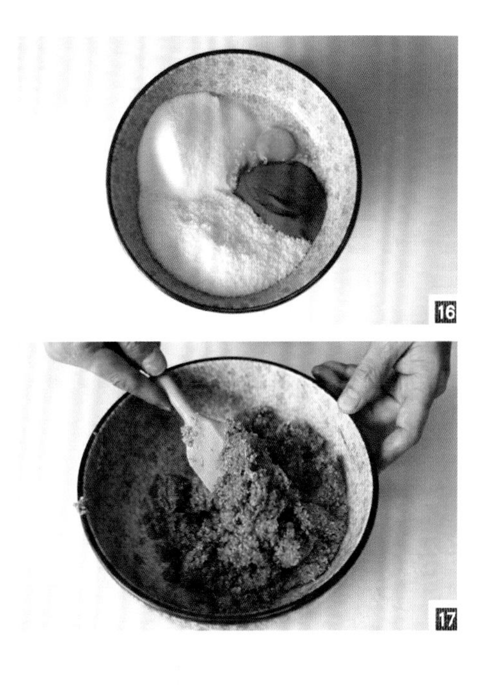

Shredded Coconut Buns / Loaves

I'm shaping the bread two ways. For the first one, cut the dough log into slices about 2 cm thick. Put each slice in a paper baking cup. They would turn out to be coconut swirl buns after baked. For the second one, cut the log across the length in half after rolling the filling in. Press each half into an 18-cm cake tin. They would turn out to be coconut loaves.

Sponge	Baker's Percentage (%)	Weight (g)
bread flour	70%	210 g
milk	43%	129 g
fresh yeast	1%	3 g
sugar	2%	6 g

Main dough		
bread flour	30%	90 g
skimmed milk powder	6%	18 g
sea salt	1.5%	4.5 g
fresh yeast	2%	6 g
milk	6%	18 g
whipping cream	6%	18 g
egg	14%	42 g
unsalted butter	10%	30 g
	191.5%	**574.5 g**

Making the sponge

Mix all sponge ingredients and knead until smooth. Transfer into a ziplock bag and refrigerate at 0 to 5°C for 8 hours.

Mixing

Put all main dough ingredients and the sponge into a mixing bowl (except butter). Mix with a rubber scraper until no dry patch is visible. (Refer to photo 1-3 in the recipe "Plain Round Rolls")

Kneading

Knead the dough until smooth. Put in unsalted butter. Refer to the photo 4-9 in the recipe "Plain Round Rolls" for method. Keep on kneading until the dough can be stretched into an elastic thin membrane without breaking. Measure the dough temperature.

First proofing

Knead the dough into a big ball. Pinch to seal the seam. Leave it at room temperature (23 to 30°C) for 30 minutes.

Punch down

Press the dough to release air bubbles. Divide into 3 pieces, weighing 140 g each. Handsquare them. Handsquare the remaining dough. Leave them to rest in a fridge for 15 to 20 minutes until the dough temperature come down to 15°C. (Refer to the recipe "Plain Round Rolls" (photo 59-61) for the method of handsquaring.)

Shaping

Buns: Roll the dough into a rectangle, measuring 30 x 10 cm. Spread the filling on. Roll into a log about 8 cm in diameter. Press to secure the seam. Cut into slices about 2 cm thick. Put them into paper baking cups. (pics. 1 - 7)

Loaves: Roll the 140 g dough out into a rectangle measuring 30 x 10 cm. Spread the filling on. Roll into a log. Cut in half across the length. Put each dough into a 18 cm cake tin greased with butter. (pics. 8 - 15)

Final proofing

Leave them for 20 to 25 minutes. Brush on egg wash.

Baking

- **Buns:** Bake in a preheated oven at 180°C for 15 minutes.
- **Loaves:** Bake in a preheated oven at 180°C for 24 minutes.

Coconut filling

Ingredients

150 g dried shredded coconut
125 g sugar
70 g egg
50 g melted butter
8 g instant coffee (optional)

Method

Mix all ingredients. (pics. 16 - 17)

紅莓合桃包

這液種法添加了粘米粉，
使成品有點 Q 軟口感。

液種材料	百分比 (%)	重量 (克)
高筋麵粉	20%	40 克
粘米粉	5%	10 克
鮮酵母	0.5%	1 克
水	25%	50 克

麵糰材料		
高筋麵粉	80%	160 克
鮮酵母	2.5%	5 克
海鹽	2%	4 克
黑糖	15%	30 克
脫脂奶粉	3%	6 克
水	45%	90 克
無鹽牛油	8%	16 克
紅莓乾	25%	50 克
合桃	25%	50 克
	256%	**512 克**

預備液種：

- 液種材料放入碗內攪拌均勻，蓋上保鮮膜，放入雪櫃發酵約 8-12 小時。（圖 1）
- 紅莓乾用熱水或酒（可選橙酒或紅酒）浸軟，隔去水份；合桃用低溫焗爐（120-130℃）烘香，掰開成大粒狀。
- 黑糖用麵糰食譜內水份浸溶或煮溶後放涼使用。

1

混合：先將鮮酵母和少許麵粉揉合，在大碗內把黑糖水注入麵粉、脫脂奶粉、鹽內，以膠刮混合至不見到水份。（參考第 38 頁「圓形小餐包」的混合步驟圖 1-3）。

搓揉：搓揉至麵糰可以投油，繼續將麵糰搓至有薄膜，量度麵糰溫度，把紅莓乾、合桃用切入法混在麵糰內，搓揉均勻。（圖 2 - 4）

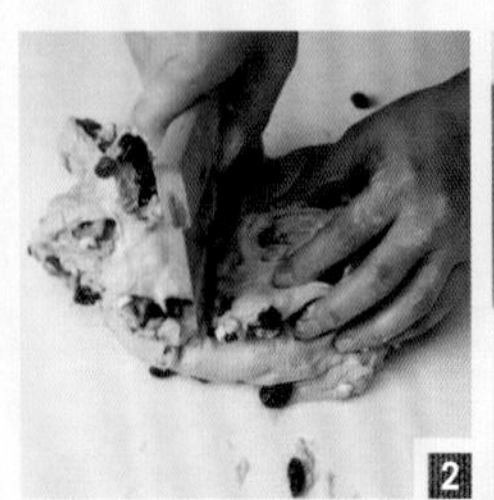
2

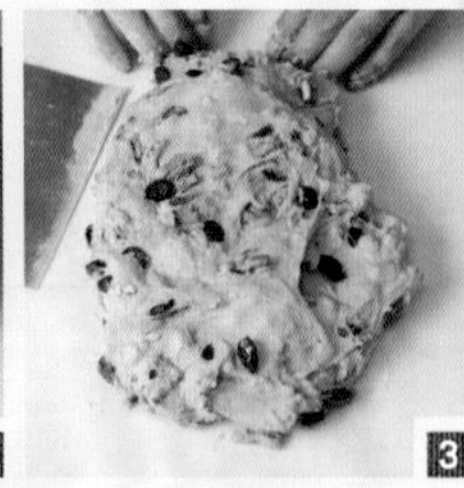
3

4

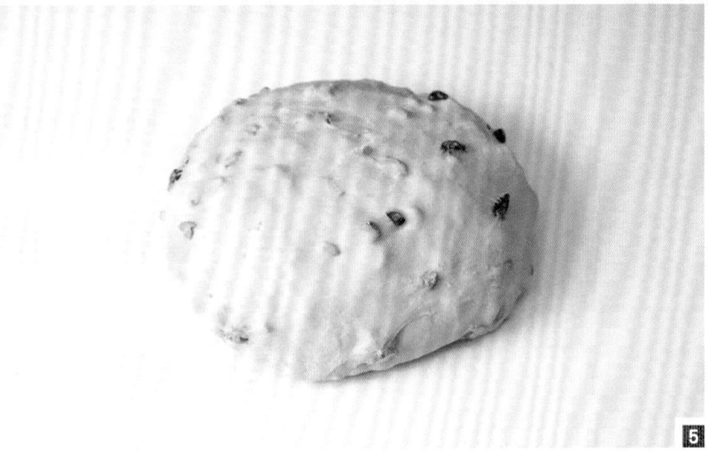

5

發酵：將搓好麵糰滾成大圓球狀，收口捏好（圖5），放室溫（23-30℃）發酵約 45-60 分鐘；測試麵糰。

排氣、分割、滾圓：第一次發酵完成，把麵糰刮出來，切成長條形再分切成 3 份（按麵糰總重量平均除 3，不可能是整數，將分剩的麵糰平分到切好的麵糰便可）。

把麵糰滾圓，放膠盒內，用蓋蓋着，讓麵糰鬆弛約 15 分鐘。

造型：桌上撒少許手粉（用高筋麵粉），把麵糰輕輕按扁，排出氣體，滾圓。

▶ **紅莓合桃包 造型 I：**把麵糰摺成橄欖形。

最後發酵：發酵約 30-45 分鐘。

剔花：在麵糰上篩米粉或高筋麵粉，用剔刀剔花。（圖 6 - 7）

烤焗：放入已預熱約攝氏 200℃焗爐，焗約 16-18 分鐘。

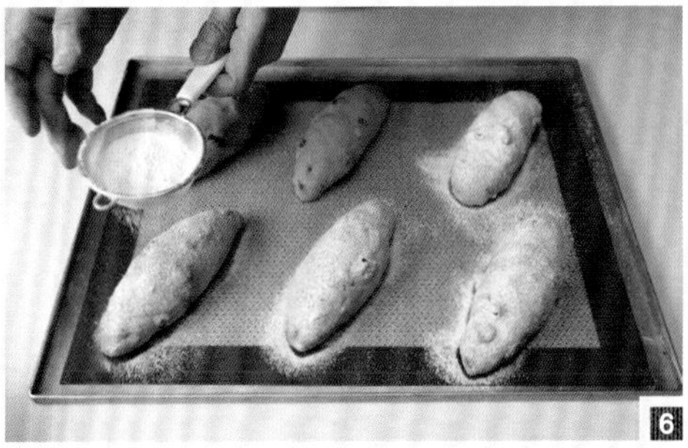

6

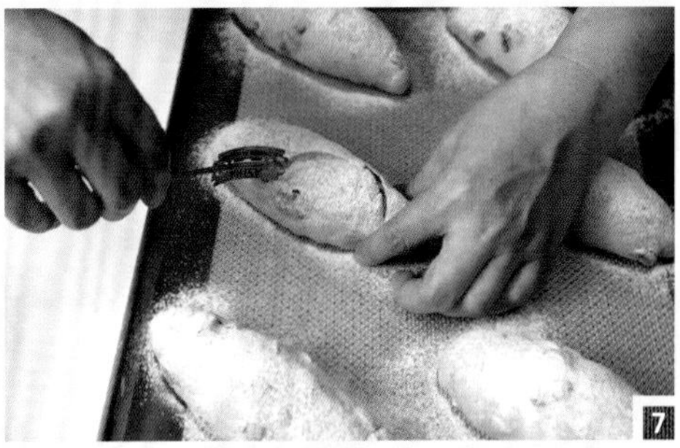

7

▶ **紅莓合桃包 造型 II**：把麵糰滾圓後接口向上，放在篩了粉的籐籃中。(圖 8 - 16)

最後發酵：大約 30-45 分鐘。

烤焗：將麵糰反扣在墊不沾布焗盤上。放入已預熱約攝氏 200℃焗爐，焗約 16-18 分鐘。

Video
紅莓合桃包
造型 II

Cranberry Walnut Bread

Poolish method

I add some rice flour to the poolish, for an extra-chewy texture.

Poolish	Baker's Percentage (%)	Weight (g)
bread flour	20%	40 g
long-grain rice flour	5%	10 g
fresh yeast	0.5%	1 g
water	25%	50 g

Main dough		
bread flour	80%	160 g
fresh yeast	2.5%	5 g
sea salt	2%	4 g
muscovado sugar	15%	30 g
skimmed milk powder	3%	6 g
water	45%	90 g
unsalted butter	8%	16 g
dried cranberries	25%	50 g
shelled walnuts	25%	50 g
	256%	**512 g**

Making the poolish

- Put all poolish ingredients into a bowl. Mix well and cover in cling film. Refrigerate for 8 to 12 hours for fermentation. (pics. 1)
- Soak the dried cranberries in hot water or liqueur (orange liqueur or red wine) until soft. Drain and set aside. Toast the walnuts in an oven at low temperature until lightly browned. Break them up into chunky pieces.
- Add muscovado sugar to the water. Stir until it dissolves. Or heat it up until it dissolves and let cool.

Mixing

Mix fresh yeast with part of the bread flour first. Transfer into a mixing bowl. Add muscovado syrup, the rest of the bread flour, skimmed milk powder and sea salt. Stir with a rubber scraper till no dry patch is visible. (Refer to photo 1-3 in the recipe "Plain Round Rolls" for the mixing method.)

Kneading

Knead the dough tlll smooth. Press it flatten to add softened butter. Keep on kneading till the dough can be stretched into an elastic membrane without breaking. Measure the temperature of the dough. Add dried cranberries and walnuts to the dough. Knead to distribute them evenly in the dough. (pics. 2 - 4)

First proofing

Roll the dough into a big ball. Pinch the seam to seal well. (pics. 5) Leave it at room temperature (23 to 30°C) for 45 to 60 minutes. Do the ripe test to see if it has proofed enough.

Punch down / dividing / rolling round

After the first proofing, scrape the dough out. Cut into long strips and then into three pieces of equal weight. Weight of each piece is determined by dividing the total weight of the dough by 3. It may not be an integer. If there is any leftover dough, just divide it among the three pieces.

Roll each dough piece round and then put them into a plastic box. Cover the lid. Let them rest for 15 minutes.

Shaping

Sprinkle some bread flour on the countertop. Put the dough pieces over. Gently flatten the dough to release big air bubbles. Roll each dough round again.

- **Option 1:** Fold each dough into an oval shape.
- **Option 2:** Roll the dough round with the seam facing up. Transfer into a floured proofing basket for final proofing. (pics. 8 - 16)

Final proofing

Let them proof for 30 to 45 minutes.

▶ **Option 1:** Sprinkle rice flour or bread flour over the dough. Slash with a bread lame.(pics. 6 - 7)

▶ **Option 2:** Turn the dough out of the proofing basket into a baking tray lined with a silicone baking mat.

Baking

Bake in a preheated oven at 200°C for 16 to 18 minutes.

蔬菜包

直接法：翻面
學習添加不同材料

材料	百分比（%）	重量（克）
法包粉	100%	300 克
砂糖	3%	9 克
海鹽	1.5%	4.5 克
脫脂奶粉	3%	9 克
水	65%	195 克
鮮酵母	3%	9 克
無鹽牛油	3%	9 克
	178.5%	**535.5 克**

配料

南瓜粒（蒸熟）	25%	75 克
粟米粒（蒸熟）	20%	60 克
西蘭花粒（用鹽水灼熟）	25%	75 克
洋葱粒（炒香）	15%	45 克
紅蘿蔔粒（用鹽水灼熟）	25%	75 克
車打芝士	20%	60 克
海鹽	0.8%	2.4 克
黑椒		適量
Mozzarella 芝士（鋪面用）		適量

預備：配料切好，按指定方法煮好。（圖 1）

麵糰做法：參考第 38 頁「圓形小餐包」的混合、搓揉、投油、測試筋膜和發酵的步驟和圖 1-25。（圖 2）

混合麵糰和配料：將搓好麵糰、配料（蔬菜粒和芝士）放入膠盒內，以切刀將麵糰切細，把配料混合均勻。（圖 3 - 8）

Video
蔬菜包
切入法

* 配料切入只可以在麵糰未發酵前處理，如果麵糰發酵後會令已成筋膜的麵筋切斷，這時候只可用包入法。

發酵：麵糰發酵 45 分鐘後，將麵糰拿出來放枱面上翻面，再發酵 20 分鐘（翻面做法看第 74 頁「全麥包」的圖 5-11）。

排氣：把麵糰 Handsquare（做法看第 43 頁「圓形小餐包」的圖 59-61），休息 15-20 分鐘。

造型：輕輕按麵糰排出氣體，用手將麵糰拍薄成方形，切成大小相若的長方形，不用量重量。（圖 9 - 12）

最後發酵：大約 20-25 分鐘，再放上 Mozzarella 芝士。

烤焗：放入已預熱約 200℃焗爐，焗約 15-16 分鐘。

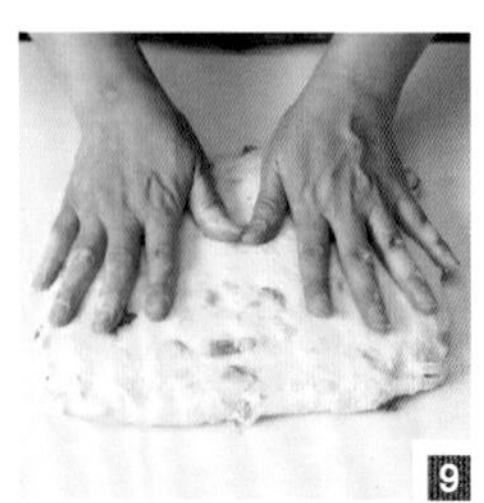
9

10

11

12

Vegetable Flat Bread

Direct method

Punching down. Learn to add extra ingredients.

Main dough	Baker's Percentage (%)	Weight (g)
French flour Type 55	100%	300 g
sugar	3%	9 g
sea salt	1.5%	4.5 g
skimmed milk powder	3%	9 g
water	65%	195 g
fresh yeast	3%	9 g
unsalted butter	3%	9 g
	178.5%	**535.5 g**

Other ingredients		
diced pumpkin (steamed)	25%	75 g
sweet corn kernels (steamed)	20%	60 g
diced broccoli (blanched in salted water)	25%	75 g
diced onion (stir-fried in oil)	15%	45 g
diced carrot (blanched in salted water)	25%	75 g
cheddar cheese (diced)	20%	60 g
sea salt	0.8%	2.4 g
ground black pepper		
grated mozzarella cheese (garnish)		

Preparation

Dice and cook the ingredients in indicated above. Let cool. (pics. 1)

Making the dough

Follow the steps of the recipe "Plain Round Rolls" for mixing, kneading, adding butter, windowpane test and first proofing.
(pics. 2)

Adding veggies to the dough

Put the dough, all veggies and cheddar cheese into a plastic box. Cut the dough finely with a knife. Knead to distribute the veggies evenly in the dough. (pics. 3 - 8)

* This method only works before the dough has proofed. It's because cutting the dough would break the gluten structure developed during fermentation. For dough that has proofed, you should wrap the veggies in as if they are filling.

First proofing

Leave the dough to proof for 45 minutes. Punching down the dough on a countertop and let it proof for 20 more minutes. (Refer to photo 5-11 in the recipe "Wholemeal loaf".

Punch down

Handsquare the dough. Leave it for 15 to 20 minutes.

Shaping

Gently press the dough to release air bubbles. Pat it with your hands to shape into a square. Then cut into rectangular pieces about the same sizes. You don't need to weight them. (pics. 9 - 12)

Final proofing

Leave them for 20 to 25 minutes. Top with mozzarella cheese.

Baking

Bake in a preheated oven at 200°C for 15 to 16 minutes.

MILK

提子包

直接法：翻面
學習添加不同材料

材料	百分比（%）	重量（克）
高筋麵粉（蛋白質：14.2%）	80%	240 克
高筋麵粉（蛋白質：11.8%）	20%	60 克
砂糖	8%	24 克
海鹽	2%	6 克
脫脂奶粉	3%	9 克
蛋黃	10%	30 克
鮮酵母	4%	12 克
無鹽牛油	10%	30 克
水	60%	180 克
提子乾	20%	60 克
金提子乾	20%	60 克
橙酒	5%	15 克
糖漬橙皮粒	15%	45 克
	257%	**771 克**

肉桂砂糖	
肉桂粉	3-5 克
砂糖	100 克

預備：將乾果用橙酒浸泡約最少 30 分鐘或過夜，瀝乾水份候用。

麵糰做法：參考第 38 頁「圓形小餐包」的混合、搓揉、投油、測試筋膜、發酵步驟和圖 1-25。麵糰搓好後，將乾果加在麵糰內，然後搓揉均勻。

麵糰發酵 80 分鐘後，將麵糰拿出來放枱面上翻面後再發酵 40 分鐘（翻面做法看第 74 頁「全麥包」的圖 5-11）。

排氣：排氣，分割成 80 克麵糰，滾圓後（按第 41 頁「圓形小餐包」的排氣、分割、滾圓的做法和圖 30-38）放室溫休息 15-20 分鐘。

造型：滾圓後放模內。（圖 1）

1

最後發酵：大約 30-40 分鐘，塗蛋，用剪刀在麵糰剪十字，灑上肉桂砂糖（視乎喜好調節肉桂粉份量）。（圖 2 - 3）

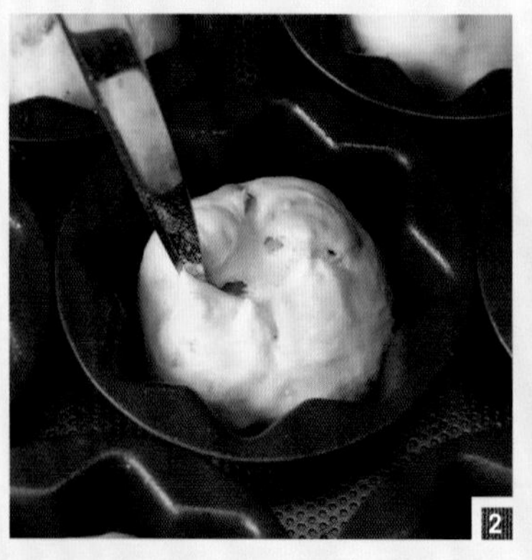
2

3

烤焗：放入已預熱約 180℃焗爐，焗約 16 分鐘。

Raisin Buns

Direct method

Punching down.

Learn to add extra ingredients.

Main dough	Baker's Percentage (%)	Weight (g)
bread flour (14.2% gluten content)	80%	240 g
bread flour (11.8% gluten content)	20%	60 g
sugar	8%	24 g
sea salt	2%	6 g
skimmed milk powder	3%	9 g
egg yolk	10%	30 g
fresh yeast	4%	12 g
unsalted butter	10%	30 g
water	60%	180 g
raisins	20%	60 g
golden sultanas	20%	60 g
orange liqueur	5%	15 g
diced candied orange peel	15%	45 g
	257%	**771 g**

Cinnamon sugar

3-5 g ground cinnamon

100 g sugar

Preparation

Soak raisins, sultanas and candied orange peel in the orange liqueur for at least 30 minutes (or overnight). Drain and set aside.

Making the dough

Follow the steps in the recipe "Plain Round Rolls" for mixing, kneading, adding butter, windowpane test and first proofing. Add raisins, sultanas and candied orange peel to the dough. Knead to distribute evenly.

Second proofing

After proofing for 80 minutes, punching down the dough on a countertop and let it proof for 40 more minutes. (Refer to photo 5-11 in the recipe "Wholemeal loaf".)

Punch down

Press the dough to release air bubbles. Divide the dough into pieces weighing 80 g each. Roll each round. (Refer to the recipe "Plain Round Rolls" for the steps on punch down, dividing, and rolling round.) Leave them at room temperature to rest for 15 to 20 minutes.

Shaping

Roll each dough round again. Press into greased tins. (pics. 1)

Final proofing

Leave them for 30 to 40 minutes. Brush on egg wash. Make a crisscross cut on the dough. Sprinkle with cinnamon sugar (feel free to adjust the amount of ground cinnamon used). (pics. 2 - 3)

Baking

Bake in a preheated oven at 180°C for 16 minutes.

大薩芝士餡包 & 吞拿魚沙律餡包

老麵法：包餡

* 老麵材料	百分比（%）	重量（克）
高筋麵粉	100%	100 克
水	65%	65 克
鮮酵母	1%	1 克
海鹽	1%	1 克

麵糰材料		
高筋麵粉	100%	250 克
海鹽	2%	5 克
脫脂奶粉	3%	7.5 克
砂糖	15%	37.5 克
蛋	10%	25 克
鮮酵母	3%	7.5 克
水	55%	137.5 克
* 老麵	20%	50 克
無鹽牛油	8%	20 克
	216%	**540 克**

餡料	
大薩芝士（切細粒）	每個包約 20 克
吞拿魚沙律餡	每個包約 40 克

老麵法：學習包餡

- ▶ 包餡的餡料我們要預先預備好，例如乾餡要切細，濕餡料要瀝乾水份，需要調味的，要先調好味。生肉餡料要預先煮熟，放涼才可以包餡，因為麵糰入爐時間很短，生肉未必能熟透。所有餡料都要分成等份。餡料的溫度亦要和麵糰溫度相若，太熱會令麵糰發酵過度，過冷就會令發酵時間太久。

- ▶ 包餡時，麵糰排氣後，用手或擀麵棍將麵糰周邊 1 厘米位置按壓得比較薄，麵糰中間比較厚，當餡料放中間位，周邊麵糰被拉起向上面集中收口時，收口的麵糰會較厚，如果麵皮開得太大或是中間位置沒有按得較厚，收在上面的麵糰就會過多，麵包就會底部過薄，變得上下厚薄不均。麵糰薄的地方會有機會爆餡，貼着餡的麵皮部份亦很難上色，變得好像未熟的樣子。

- ▶ 包餡後麵糰就不要再滾圓，否則餡料會走位。

- ▶ 乾餡料如芝士、肉鬆等等沒有汁水比較容易包，只注意切得比較細小就不會弄穿麵糰。

按老麵做法預備老麵（看第 28 頁「造麵包的方法」）。

混合：先將鮮酵母溶於食譜份量的部份水中或灑在麵粉內，再把酵母水放入麵粉、餘下水、糖、蛋、脱脂奶粉、老麵和鹽，在大碗內以膠刮混合至不見到水份。

搓揉：搓揉至麵糰光滑可以投油，按第 38 頁「圓形小餐包」步驟和圖 4-19，繼續搓麵糰至有薄膜，量度麵糰溫度。

發酵：將搓好麵糰滾成大圓球狀，收口捏好，放室溫（23-30℃）發酵約 45-60 分鐘；測試麵糰。

排氣、分割：第一次發酵完成，把麵糰刮出來，弄成方形，切成長條形再切細成大約 65 克一個（按麵糰總重量平均除 8，不可能是整數，將分剩的麵糰平分到切好麵糰便可）。

滾圓：把麵糰滾圓，放膠盒內，用蓋蓋着，讓麵糰鬆弛約 15 分鐘。

造型：桌上撒少許手粉（用高筋麵粉），把麵糰輕輕按扁，排出氣體，圓形和三角形周邊 1 厘米位置按壓得比較薄，橄欖形造型只需將麵糰拍成橢圓形。包入芝士（圖 1 - 14）或吞拿魚沙律餡（圖 15 - 26），放在焗盤上作最後發酵。

小圓餐包芝士餡造型圖（圖 1 - 6）

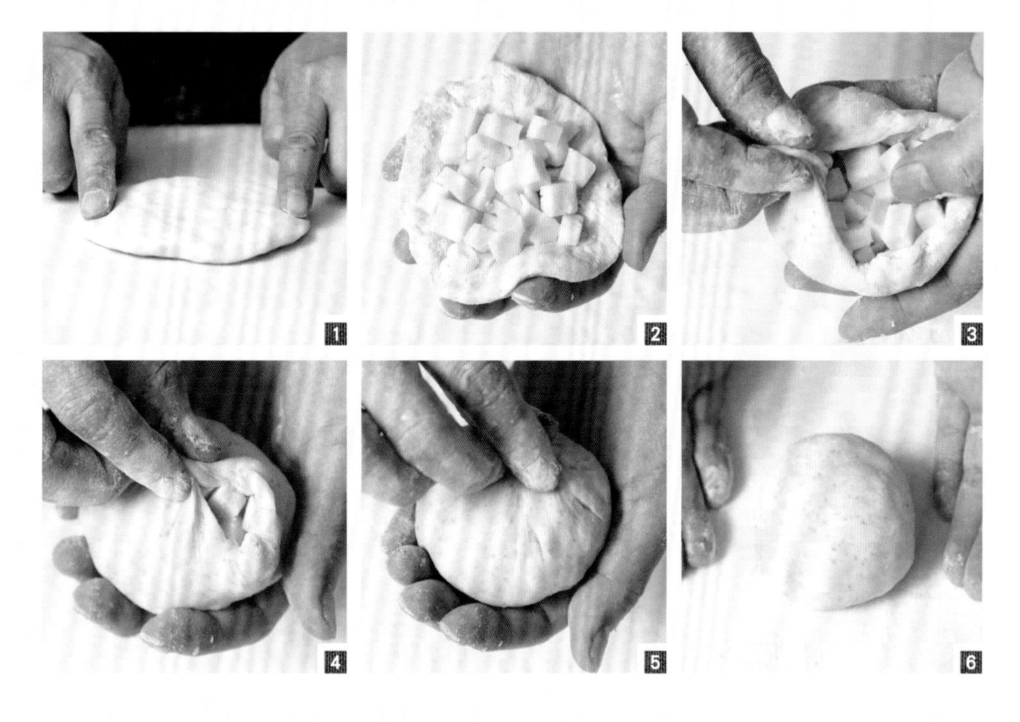

橄欖形包芝士餡造型圖（圖 7 - 9）

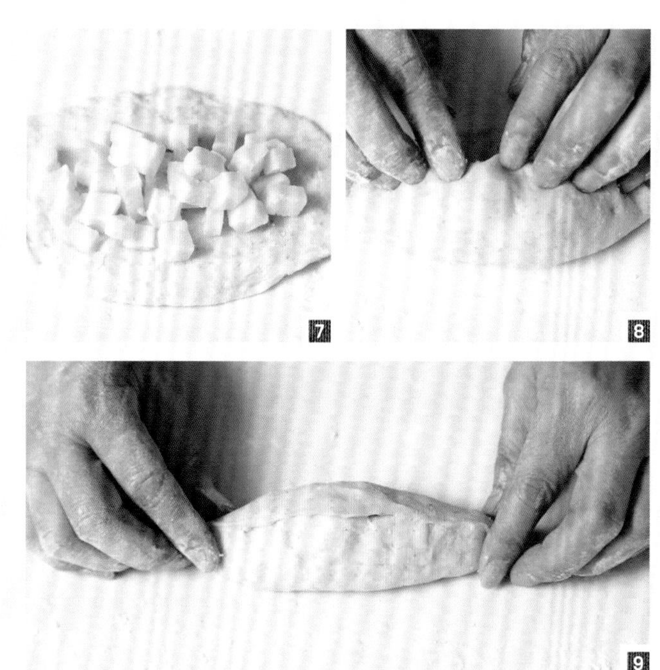

Video
小圓餐包芝士餡造型

三角形包芝士餡造型圖（圖 10 - 14）

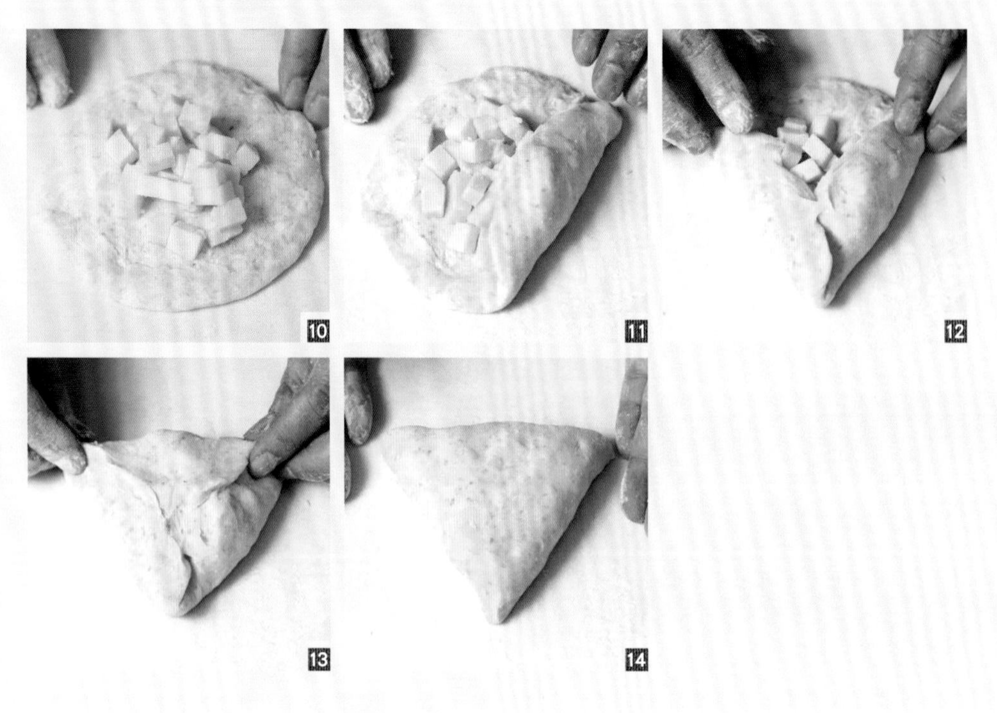

小圓餐包吞拿魚沙律餡造型圖（圖 15 - 20）

橄欖形包吞拿魚沙律餡造型圖（圖 21 - 26）

最後發酵：大約發酵 30-40 分鐘，入爐前裝飾。

烤焗：放入已預熱約 180-200℃焗爐，焗約 13-14 分鐘。

吞拿魚沙律餡製作方法

材料

罐頭吞拿魚	300 克
罐頭粟米粒	80 克
沙律醬	50 克
海鹽、黑胡椒	適量
雜芝士碎（薄餅芝士或用車打芝士）	50 克

做法

罐頭吞拿魚拆碎和罐頭粟米粒、沙律醬、海鹽、胡椒、雜芝士碎混合。

Cheese Rolls / Tuna Rolls

Old dough method： Learn about wrap filling

- For any buns or rolls with filling, you should make the filling in advance. Dry filling ingredients should be finely diced. Wet filling ingredients should be drained. All seasoning should be done beforehand. Any raw meat that you use as filling should be pre-cooked and let cool before wrapped in the dough. It's because the meat may not cook as quickly as the raw dough in the oven, so that the meat filling may not be cooked through when the bread is baked. Filling should be divided into equal portions and it should be about the same temperature as the dough. If the filling is too warm, the dough may be over-proofed. If the filling is too cold, it needs too much time to proof.

- After punching down, press the dough with your fingers to make a 1 cm wide edge that it is considerably thinner than the centre. Or, you can do it with a rolling pin. The filling will be put at the centre and the edges will be gathered to seal in the filling. The gathered seam tends to be thicker than the rest of the bun. If the dough is rolled out too much, or if the centre is not thicker than the edges, the gathered seam will be too thick while the rest of the bun will be too thin. The weak spots on the bun may burst when baked. If the filling to too close to the top of the bun, that spot tends to brown very slowly. The bun may look as if it's not baked enough.

- After you wrap filling in the dough, do not roll it round again. Otherwise, the filling may shift its position instead of staying at the centre of the bun.

- Dry fillings such as cheese or dried pork floss are easier to handle as they don't give much moisture. Just make sure you dice them finely so that they won't poke through the dough.

*Old dough	Baker's Percentage (%)	Weight (g)
bread flour	100%	100 g
water	65%	65 g
fresh yeast	1%	1 g
sea salt	1%	1 g

Main dough		
bread flour	100%	250 g
sea salt	2%	5 g
skimmed milk powder	3%	7.5 g
sugar	15%	37.5 g
egg	10%	25 g
fresh yeast	3%	7.5 g
water	55%	137.5 g
*old dough	20%	50 g
unsalted butter	8%	20 g
	216%	**540 g**

Filling	
finely diced Swiss cheese	20 g per roll
tuna and corn salad	40 g per roll

Old dough

Make old dough according to the recipe on p.55-56 under "Different ways of breadmaking".

Mixing

Dissolves the fresh yeast in part of the water. (Or sprinkle it in the flour.) In a mixing bowl, put in flour and add the yeast mixture. Add the remaining water, sugar, egg, milk powder, old dough and salt (except butter and filling). Mix with a rubber scraper until no dry patch is visible.

Kneading

Knead the dough until smooth and add softened butter. Follow the steps in the recipe "Plain Round Rolls" for kneading and adding butter. Keep on kneading until the dough can be stretched into a thin elastic membrane. Check dough temperature.

First proofing

Roll the dough into a big ball. Pinch to seal the seam. Leave it at room temperature between 23 to 30°C for 45 to 60 minutes. Do the ripe test to check if it has risen enough.

Punch down / dividing

After the first proofing, scrape out the dough. Shape it into a square. Cut into long strips and then into small pieces about 65 g each. (The weight of each dough piece is determined by dividing the total weight of the dough by 8. It may not be an integer. If there is any leftover, just divide it evenly among all dough pieces.)

Rolling round

Roll each dough piece round and put them into a plastic box. Cover the lid and let them rest for 15 minutes.

Cheese Rolls

Shaping

Sprinkle some bread flour on the countertop. Gently press each dough piece to release the big air bubbles. If you want to make round or triangular rolls, press to make a 1 cm wide edge which is considerably thinner than the centre. For oval rolls, just shape it like a rugby ball. Stuff it with cheese (pic. 1 - 14) or tuna and corn salad (15 - 26). Arrange on a baking tray for final proofing.

Final proofing

Let the dough proof for 30 to 40 minutes. Decorate before baking.

Baking

Bake in a preheated oven at 180 to 200°C for 13 to 14 minutes. Remove from oven and let cool on cooling rack.

Tuna and corn salad filling

Ingredients

canned tuna	300 g
corn kernels	80 g
salad dressing	50 g
sea salt	1 g
black pepper	
assorted grated cheeses (pizza topping or cheddar cheese)	50 g

Method

Shred tuna. Mix all the ingredients well.

Tuna Rolls

白汁雞包

直接法：翻面

學習處理熟餡材料

麵糰材料	百分比（%）	重量（克）
高筋麵粉	90%	270 克
全麥粉	10%	30 克
砂糖	10%	30 克
海鹽	2%	6 克
脫脂奶粉	3%	9 克
蛋	10%	30 克
蛋黃	8%	24 克
鮮酵母	3%	9 克
水	55%	165 克
無鹽牛油	12%	36 克
	203%	**609 克**

餡料
白汁雞肉

EPICUREAN

麵糰做法：參考第 38 頁「圓形小餐包」的混合、搓揉、投油、測試筋膜和發酵的步驟和圖 1-25。

發酵：麵糰發酵 60 分鐘後，將麵糰拿出來放枱面上翻面，再發酵 30 分鐘。（翻面做法看第 74 頁「全麥包」的圖 5-11）。

排氣：排氣，分割成 65 克麵糰滾圓，放雪櫃休息 20-25 分鐘至麵糰大約 15℃。

造型：排氣、滾圓，擀薄約 1.5 厘米，放入塗油模中。（圖 1 - 2）

最後發酵：發酵大約 25 分鐘，至鬆身，將白汁雞肉舀在麵糰上，再放上 Mozzarella 芝士。（圖 3 - 5）

烤焗：放入已預熱約 180-200℃焗爐，焗約 16 分鐘。（圖 6）

3

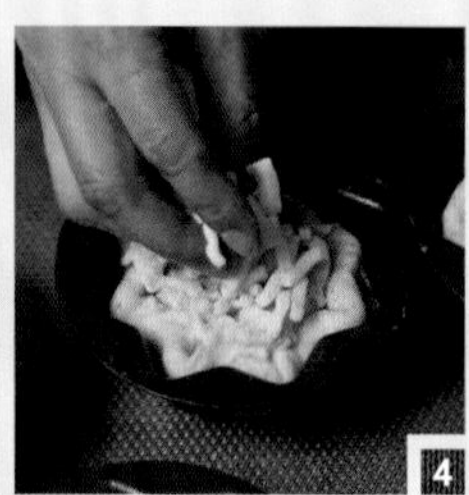
4

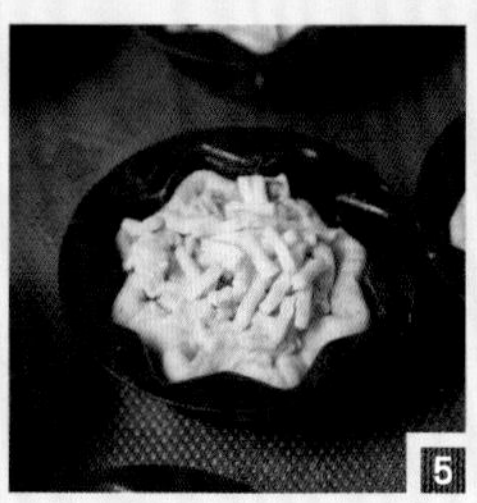
5

6

白汁雞肉製作方法

材料		材料	
無鹽牛油	30 克	蒜茸	適量
淡忌廉	50 克	雞髀肉	350 克
奶	200 克	洋葱粒	120 克
香葉	1 片	車打芝士碎	100 克
低筋麵粉	25 克	蘑菇	100 克
海鹽	3 克	鹽、黑椒、粟粉	適量
橄欖油	適量		

白汁雞肉做法

將無鹽牛油、香葉加入鍋中，以小火煮溶牛油，棄去香葉，小心不要燒焦；熄火，將低筋麵粉全部倒進鍋內，以打蛋器攪拌至完全融合。（圖 7 - 10）

兩者融合後，開中小火一直攪拌，出現泡泡後，便逐少加入奶，一邊加入一邊用打蛋器攪拌。（圖 11 - 12）

一直攪拌至牛奶滾起見泡泡和變滑便成，最後加入適量鹽和黑椒來調味。（圖 13）

雞髀肉、洋葱切丁，用少許鹽、黑椒、粟粉略醃，蘑菇切片。

以橄欖油和蒜茸起鍋，爆香洋葱、雞髀肉，加入蘑菇炒香，拌入白汁，放涼後拌入芝士碎。（圖 14 - 18）

Chicken a la King Buns

Direct method: Punching down

Learning about adding ingredients on top of bread.

Main dough	Baker's Percentage (%)	Weight (g)
bread flour	90%	270 g
wholemeal flour	10%	30 g
sugar	10%	30 g
sea salt	2%	6 g
skimmed milk powder	3%	9 g
egg	10%	30 g
egg yolk	8%	24 g
fresh yeast	3%	9 g
water	55%	165 g
unsalted butter	12%	36 g
	203%	**609 g**

Creamy chicken topping

Ingredients		Ingredients	
unsalted butter	30 g	grated garlic	
whipping cream	50 g	boneless chicken thigh	350 g
milk	200 g	diced onion	120 g
bay leaf	1 piece	diced cheddar cheese	100 g
cake flour	25 g	button mushrooms	100 g
sea salt	3 g	salt, black pepper and cornstarch	
olive oil			

Making the dough

Refer to the recipe "Plain Round Rolls" for mixing, kneading, adding butter, windowpane test and first proofing.

Second proofing

After proofing for 60 minutes, punching down the dough on the countertop. Let it rise for 30 more minutes. (Refer to photo 5-11 in the recipe "Wholemeal loaf".)

Punch down

Press the dough to release air bubbles. Then divide into dough pieces about 65 g each. Roll them round. Leave them to rest in a fridge for 20 to 25 minutes until the dough cools down to 15°C.

Shaping

Punch down and roll the dough pieces round again. Roll each out into 1.5-cm thick pieces with a rolling pin. Press into a greased tin. (pics. 1 - 2)

Final proofing

Leave them to proof for 25 minutes until fluffy. Scoop the creamy chicken topping over the dough. Sprinkled with grated mozzarella. (pics. 3 - 5)

Bake in a preheated oven at 180 to 200°C for 16 minutes. (pics. 6)

Method

1. Heat unsalted butter in a pan over low heat. Add a bay leaf. Be careful not to burn the butter. Discard the bay leaf. Turn off the heat and stir in the cake flour. Whisk until well incorporated. (pics. 7 - 10)
2. Turn on medium-low heat while stirring the mixture until it bubbles. Slowly add in milk while stirring continuously. (pics. 11 - 12)
3. When the mixture starts to bubble and turns creamy, season with salt and black pepper. This is the Béchamel sauce. (pics. 13)
4. Dice the chicken thigh and onion. Add a pinch of salt, black pepper and cornstarch and mix well. Set aside. Slice the mushrooms.
5. Heat a pan and add olive oil and grated garlic. Stir-fry onion and chicken thigh until lightly browned. Add mushrooms and stir well. Pour in the Béchamel sauce and mix well. Leave it to cool. Stir in diced cheddar. (pics. 14 - 18)

朱古力吉士包

直接法：學習添加或轉換粉類材料

麵糰材料	百分比（%）	重量（克）
高筋麵粉（蛋白質約 14%）	80%	200 克
高筋麵粉（蛋白質約 11.5%）	20%	50 克
可可粉	6%	15 克
砂糖	25%	62.5 克
海鹽	1.5%	3.75 克
蛋黃	8%	20 克
脫脂奶粉	3%	7.5 克
鮮酵母	3%	7.5 克
水	50%	125 克
淡忌廉	16%	40 克
無鹽牛油	15%	37.5 克
	227.5%	**568.75 克**

麵糰做法：參考第 38 頁「圓形小餐包」的混合、搓揉、投油、測試筋膜、發酵、測試、排氣、分割和滾圓的步驟和圖 1-38。

造型：把麵糰輕輕按扁，排出氣體，用擀麵棍將麵糰擀薄成橢圓形，唧約 30 克橙吉士醬在中間，把麵皮覆過來要短於下面麵皮，這樣才不易爆開，按緊收口，反轉把底部朝上，在邊位處平均切三刀約 2 厘米深，放在焗盤上作最後發酵。（圖 1 - 7）

最後發酵：大約發酵 25-30 分鐘，塗上蛋液。

烤焗：放入已預熱約 160-180℃焗爐，焗約 15 分鐘。

深色麵糰不宜太高火烤焗，會容易烤焦。

橙吉士醬製作方法

材料

奶	250 克
雲呢拿枝或天然雲呢拿油	半枝 /1 茶匙
蛋黃	70 克
砂糖	70 克
低筋麵粉	30 克
半糖漬橙皮或糖漬橙皮或橙果醬	60 克

做法

砂糖和蛋黃攪拌均勻，加入已篩的低筋麵粉，輕輕攪勻。(圖 8 - 9)

奶和雲呢拿枝煮滾後，先放少許在蛋黃糖內攪勻，再倒入其餘奶，攪勻。(圖 10)

蛋黃奶過篩，再煮滾，煮時要不停攪拌以免煮焦。(圖 11 - 12)

倒入一平盤內，加入半糖漬橙皮或糖漬橙皮，拌勻，貼面放上保鮮紙，立即冷卻，放入即棄唧袋或分成 30 克一個。(圖 13 - 14)

* 如選用橙果醬，可減少砂糖。

Cocoa Buns with Orange Custard Filling

Direct method

Learn about adding or converting dry ingredients.

Main dough	Baker's Percentage (%)	Weight (g)
bread flour (about 14% gluten)	80%	200 g
bread flour (about 11.5% gluten)	20%	50 g
cocoa powder	6%	15 g
sugar	25%	62.5 g
sea salt	1.5%	3.75 g
egg yolk	8%	20 g
skimmed milk powder	3%	7.5 g
fresh yeast	3%	7.5 g
water	50%	125 g
whipping cream	16%	40 g
unsalted butter	15%	37.5 g
	227.5%	**568.75 g**

Dough

Follow the steps in the recipe "Plain Round Rolls" for mixing, kneading, adding butter, windowpane test, first proofing, ripe test, punch down, dividing and rolling round.

Shaping

Gently press each dough piece flat to release air bubbles. Roll each dough piece out into an oval shape. Pipe 30 g of orange custard at the centre. Fold the dough along the length once so that the top dough is slightly shorter than the bottom dough. The bun is less likely to burst that way. Press to seal the seam. Flip the bun upside down. Make 3 cuts about 2 cm deep all the way through across the rounded edge. Transfer onto a baking tray for proofing. (pics. 1 - 7)

Final proofing

Leave them for 25 to 30 minutes. Brush egg wash on top.

Baking

Bake in a preheated oven at 160 to 180°C for about 15 minutes. Darker dough should not be baked at a temperature too high as it tends to brown more easily.

Orange custard

Ingredients	Weight (g)
milk	250 g
vanilla pod	1/2
(or 1 tsp natural vanilla essence)	
egg yolk	70 g
sugar	70 g
cake flour	30 g
semi-candied orange peel	60 g
(or candied orange peel or orange marmalade)	

* If you use orange marmalade, you may reduce the amount of sugar used accordingly.

Method

1. Mix sugar with egg yolk until well combined. Sieve in cake flour. Fold gently. (pics. 8 - 9)
2. If you use a vanilla pod, cut open the pod and scrape out the black seeds. Put both the seeds and the pod into a pot and add milk. Bring to the boil. Pour a little of the milk into the sugar and egg yolk mixture and stir quickly to temper it. Then pour in the rest of the hot milk while stirring to mix well. (pics. 10)
3. Strain the mixture via a wire mesh back into the pot. Turn on the heat and keep stirring until it comes to a boil. That would prevent the custard from burning or turning lumpy. (pics. 11 - 12)
4. Pour the custard into a flat tray. Add semi-candied orange peel. Stir well. Cover with cling film so that the cling film touches the custard. Leave it to cool. Transfer into a piping bag (or divide into 30 g portions). (pics. 13 - 14)

紫薯花花包

老麵法：學習包餡處理

* 老麵材料	百分比（%）	重量（克）
高筋麵粉	100%	100 克
水	65%	65 克
鮮酵母	1%	1 克
海鹽	1%	1 克

麵糰材料		
高筋麵粉	100%	250 克
海鹽	2%	5 克
脫脂奶粉	3%	7.5 克
砂糖	15%	37.5 克
蛋	10%	25 克
鮮酵母	3%	7.5 克
水	55%	137.5 克
* 老麵	20%	50 克
無鹽牛油	8%	20 克
	216%	**540 克**

紫薯餡	
台灣或日本番薯粉	70 克
熱水	70-80 克
砂糖	30 克
無鹽牛油	15 克

預備：將所有紫薯餡料放大碗內，混合攪拌均勻，分成 30 克一個，放涼待用。（圖 1 - 5）

老麵：老麵做法請看第 28 頁的「造麵包的方法」。

混合：先將鮮酵母溶於食譜份量的部份水中或灑在麵粉內，再把酵母水放入麵粉、餘下水、糖、蛋、脱脂奶粉、老麵和鹽內，以膠刮混合至不見到水份。

搓揉：搓揉至麵糰光滑可以投油，按第 38 頁「圓形小餐包」步驟和圖 4-19，繼續搓麵糰至有薄膜；量度麵糰溫度。

發酵：將搓好麵糰滾成大圓球，收口揑好，放室溫（23-30℃）發酵約 45-60 分鐘；測試麵糰。

排氣、分割：第一次發酵完成，把麵糰刮出來，弄成方形，切成長條形，再切細成大約 65 克一個（按麵糰總重量平均除 8，不可能是整數，將分剩的麵糰平分到切好的麵糰便可）。

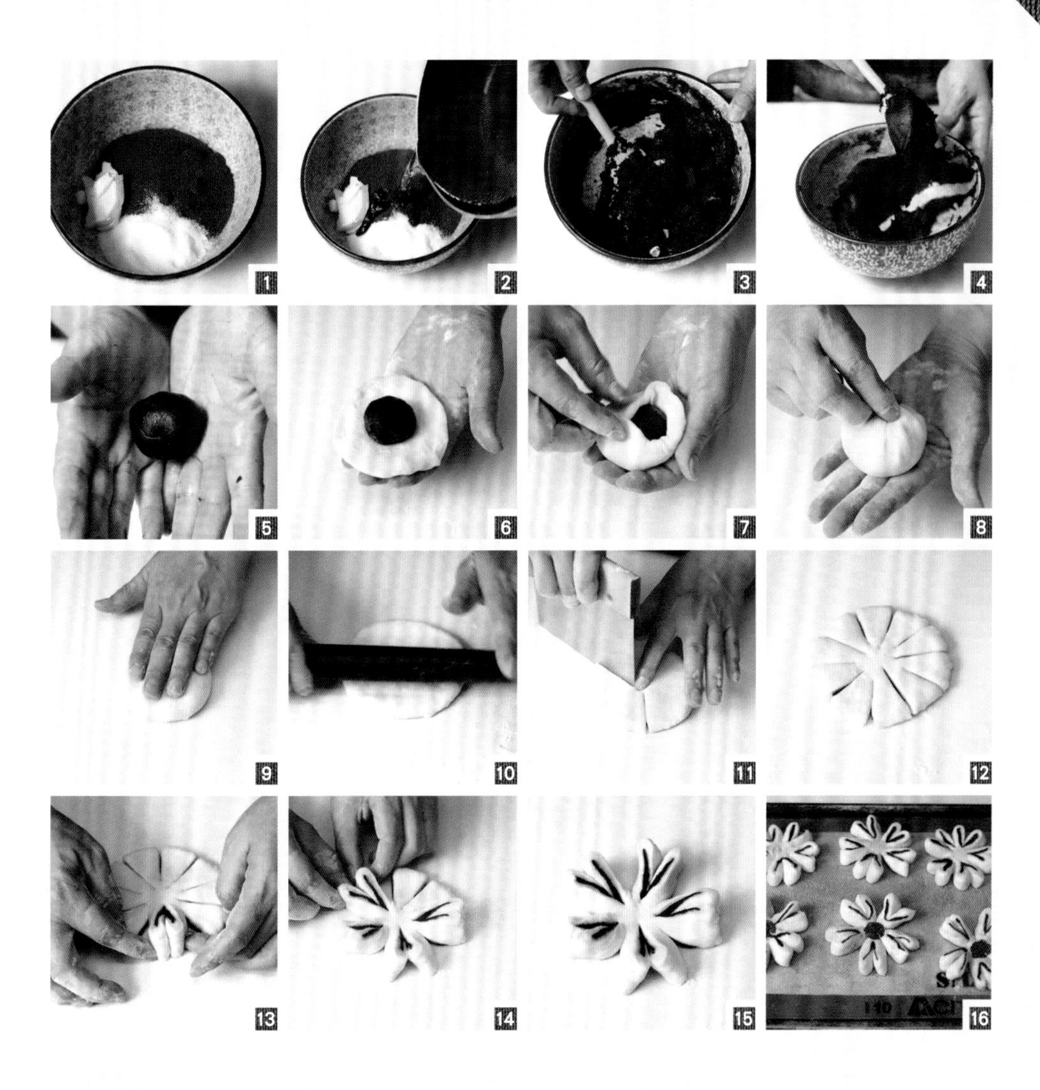

滾圓：把麵糰滾圓，放膠盒內，用蓋蓋着，讓麵糰鬆弛約 15 分鐘。

造型：桌上撒少許手粉（用高筋麵粉），先把所有麵糰輕輕按扁，排出氣體，包入紫薯餡，收緊口，滑面向上，用手按扁，再用擀麵棍擀薄約 1 厘米厚，指尖按在麵糰為中心點，平均切 10 刀到底，中心點約剩 2 厘米不切。（圖 6 - 15）

兩片花瓣對扭成心形，放在焗盤上作最後發酵。

最後發酵：壓薄了的麵糰很快發酵好，最後發酵大約只要 20 分鐘，塗蛋，裝飾。（圖 16）

烤焗：放入已預熱約 200-220℃焗爐，焗約 10 分鐘。

這麵包較薄，用高火快烤，可保持濕潤柔軟，不宜烤過度至乾硬。

Flower-Shaped Buns with Purple Sweet Potato Filling

Old dough method

Learn about wrap in filling.

*Old Dough Ingredients	Baker's Percentage (%)	Weight (g)
bread flour	100%	100 g
water	65%	65 g
fresh yeast	1%	1 g
sea salt	1%	1 g

Main dough		
bread flour	100%	250 g
sea salt	2%	5 g
skimmed milk powder	3%	7.5 g
sugar	15%	37.5 g
egg	10%	25 g
fresh yeast	3%	7.5 g
water	55%	137.5 g
*old dough	20%	50 g
unsalted butter	8%	20 g
	216%	**540 g**

Purple sweet potato filling	
Taiwanese or Japanese purple sweet potato flour	70 g
hot water	70-80 g
sugar	30 g
unsalted butter	15 g

Preparation

To make the purple sweet potato filling, put all ingredients into a mixing bowl. Stir to mix well. Divide into balls about 30 g each. Let cool. (pics. 1 - 5)

Old Dough

Make old dough according to the recipe on p.55-56 under "Different ways of breadmaking".

Mixing

Dissolve the fresh yeast in part of the water in the recipe. (Or, you may mix fresh yeast in the flour.) Put fresh yeast mixture into a mixing bowl. Add flour, remaining water, sugar, egg, salt and all remaining ingredients (except butter). Stir with a rubber scraper until no dry patch is visible.

Kneading

Knead the dough until smooth. Then add softened butter. Follow steps of the recipe "Plain Round Rolls". Keep on kneading until you can stretch the dough into a thin elastic membrane. Measure the temperature of the dough.

First proofing

Roll the dough into a big ball. Pinch the seam to seal well. Leave it at room temperature (23 to 30°C) to proof for 45 to 60 minutes. Do the ripe test to see if it has proofed enough.

Punch down / dividing

Scrape out the dough. Shape it into a square. Cut into long strips and then cut into small pieces weighing roughly 65 g each. (The weight of each dough piece is determined by dividing the total weight of the dough by 8. It may not be an integer. If there is any leftover dough, just divide it among the dough pieces evenly.)

Rolling round

Roll each dough piece round. Put them into a plastic box and cover the lid. Leave them to rest for 15 minutes.

Shaping

Sprinkle bread flour on a countertop. Gently press the dough pieces to release big air bubbles. Wrap in one ball of purple sweet potato filling. Gather the edges of the dough and pinch to seal the seams. Turn it upside down so that the seams face down. Press gently to flatten. Then roll it out evenly until 1 cm thick. Hold the bun in place by pressing your finger at the centre. Make 10 cuts to cut through across the edge of the dough evenly, saving about 2 cm at the centre uncut. (pics. 6 - 15)

Twist every two adjacent petals of dough in opposite direction to make a heart shape. Arrange on a baking tray for final proofing.

Final proofing

The dough tends to proof very quickly after rolled flat. It usually takes about 20 minutes. Brush on egg wash. Decorate if desired.(pics. 16)

Baking

Preheat an oven to 200 to 220°C. Bake the buns for 10 minutes.

Buns or bread that are rolled thin should be baked over high heat for a shorter period of time. You can keep the crumbs moist this way. Do not over-bake them. Otherwise, they may turn out dry and hard.

腸仔包

湯種法

* 湯種材料	百分比（%）	重量（克）
高筋麵粉	20%	20 克
水	100%	100 克

麵糰材料		
高筋麵粉	100%	250 克
海鹽	2%	5 克
脫脂奶粉	3%	7.5 克
砂糖	15%	37.5 克
蛋	10%	25 克
鮮酵母	3%	7.5 克
水	32%	80 克
* 湯種	30%	75 克
無鹽牛油	8%	20 克
	203%	**507.5 克**

火雞肉香腸		8 條

預備湯種：水和筋粉混合攪拌均匀，過篩，放鍋內煮至75℃，放涼待用。（圖 1 - 4）

火雞肉香腸洗淨，抹乾水分。

混合：先將鮮酵母溶於食譜份量的部份水中或灑在麵粉內，在大碗內再把酵母水放入麵粉、脱脂奶粉、餘下水、糖、蛋、湯種和鹽，以膠刮混合至不見到水份。

搓揉：搓揉至麵糰可以放入牛油，按第 38 頁「圓形小餐包」搓揉步驟和圖 4-19，繼續搓麵糰至有薄膜，量度麵糰溫度。

發酵：將搓好麵糰滾成大圓球，收口捏好，放室溫（23-30℃）發酵約 45-60 分鐘；測試麵糰。（參考第 40 頁「圓形小餐包」的發酵步驟和圖 20-27）。

排氣、分割：第一次發酵完成，把麵糰刮出來，弄成方形，切成長條形，再按麵糰總重量平均除 8，大約 63 克一個。

滾圓：把麵糰滾圓，放膠盒內，用蓋蓋着，讓麵糰鬆弛約 15 分鐘，（參考第 41 頁「圓形小餐包」的排氣、分割、滾圓步驟和圖 30-38）。

造型：桌上撒少許手粉（用高筋麵粉），把麵糰輕輕按扁，排出氣體，用擀麵棍將麵糰擀薄，上下摺成圓錐形，收口捏好，搓成腸仔三倍長的長條，在香腸頭預留約 1.5 厘米開始捲起麵糰條，捲時不要壓扁麵糰條，保持圓潤，把收口捏緊在後面，放在焗盤上作最後發酵。（圖 5 - 14）

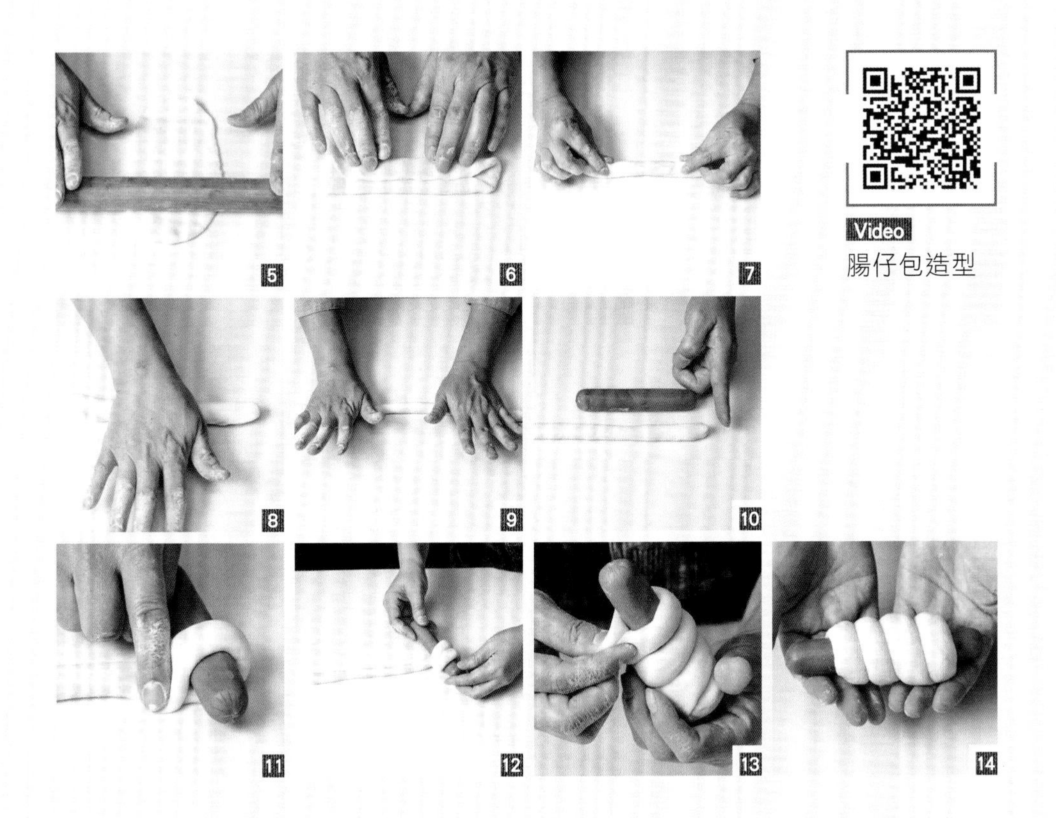

Video
腸仔包造型

最後發酵：大約發酵 20-25 分鐘，塗上蛋液，裝飾。

注意發酵時間不宜過久，以免花紋膨脹得太大而失去紋路。

烤焗：放入已預熱約 180℃焗爐，焗約 13-14 分鐘。

貼士

- 湯種法不只一種，轉用其他方法要留意水份。
- 市面一般凍肉店或超市售賣的雞肉香腸油份很高，烤焗時會溢出很多油而令周遭包着的麵包有一大圈虛位，甚至香腸也會掉下來；所以除了製作時要抹乾雞肉香腸的水分外，購買質量好的火雞肉香腸是較佳的選擇。
- 也不應使用油份多的生肉腸，如必須使用，請先將生肉腸用水煮熟才包入麵糰內。

Sausage Rolls

Yudane (Utane) method

*Yudane	Baker's Percentage (%)	Weight (g)
bread flour	20%	20 g
water	100%	100 g
Main dough		
bread flour	100%	250 g
sea salt	2%	5 g
skimmed milk powder	3%	7.5 g
sugar	15%	37.5 g
egg	10%	25 g
fresh yeast	3%	7.5 g
water	32%	80 g
*Yudane	30%	75 g
unsalted butter	8%	20 g
	203%	**507.5 g**

8 turkey franks

Tips

- There are different recipes for Yudane. If you use a different recipe, make sure you pay attention to the amount of water used.
- Chicken franks from frozen deli stores and supermarket are usually high in fat content. When baked, they tend to give much oil, making them separate from the bread. Sometimes the chicken franks may even fall off. Thus, before coiling the dough around the franks, make sure you wipe them dry. Besides, it also makes sense to get turkey franks of better quality.
- For this recipe, do not use sausages with raw meat inside because of their high fat content. If you don't have other choices, blanch them in boiling water until cooked through, let cool and wipe them dry before use.

Preparation

To make the Yudane, mix water with bread flour and stir well. Pass the mixture through a wire mesh into a pot. Heat the mixture up to 75°C. Set aside to let cool. (pics. 1 - 4) Rinse the turkey franks. Wipe dry.

Mixing

Dissolve fresh yeast in part of the water specified in the recipe. (Or sprinkle fresh yeast in the flour.) Transfer the fresh yeast mixture into a mixing bowl. Add the remaining water, sugar, egg, flour, skimmed milk powder, Yudane and salt. Mix with a rubber scraper until no dry patch is visible.

Kneading

Knead until the dough is smooth. Add butter. Follow the steps of the recipe "Plain Round Rolls". Keep on kneading until the dough can be stretched into a thin elastic membrane. Measure the temperature of the dough.

First proofing

Roll the dough into a big ball. Pinch the seam to seal well. Leave it at room temperature (23 to 30°C) for 45 to 60 minutes. Do the ripe test to see if it has risen enough.

Punch down / dividing

After the first proofing, scrape out the dough. Shape into a square. Cut into long strips and then into pieces. The weight of each piece is determined by dividing the total weight of the dough by 8, about 63 g each.

Rolling round

Roll each dough piece round. Put them into a plastic box and cover the lid. Leave them to rest for 15 minutes.

Shaping

Sprinkle some bread flour on the countertop. Gently press each dough ball to release big air bubbles. Roll it out with a rolling pin. Fold the longer ends so that they are tapered. Pinch the seam. Then roll the dough into a log with tapered ends about three times the length of a turkey frank. Coil the dough around a turkey frank leaving 1.5 cm of each end uncovered. When you coil the dough, try to keep it round in cross section. Tuck the end of the dough on the bottom. Arrange on a baking tray for final proofing. (pics. 5 - 14)

Final proofing

Leave them to proof for 20 to 25 minutes. Brush on egg wash. Decorate as desired.

Make sure you do not over-proof the rolls. Otherwise, the dough may expand too much so that the coiling folds will be less obvious.

Baking

Bake in a preheated oven at 180°C for 13 to 14 minutes.

硬小餐包

法國老麵法：

小法包、麥穗包、雙胞胎包

* 發酵種材料	百分比（%）	重量（克）
法包粉	100%	100 克
海鹽	2%	2 克
麥芽精	0.5%	0.5 克
鮮酵母	0.5%	0.5 克
水	65%	65 克

麵糰材料		
法包粉	100%	300 克
麥芽精	0.5%	1.5 克
海鹽	2%	6 克
鮮酵母	2%	6 克
水	66%	198 克
* 發酵種	15%	45 克
	185.5%	**556.5 克**

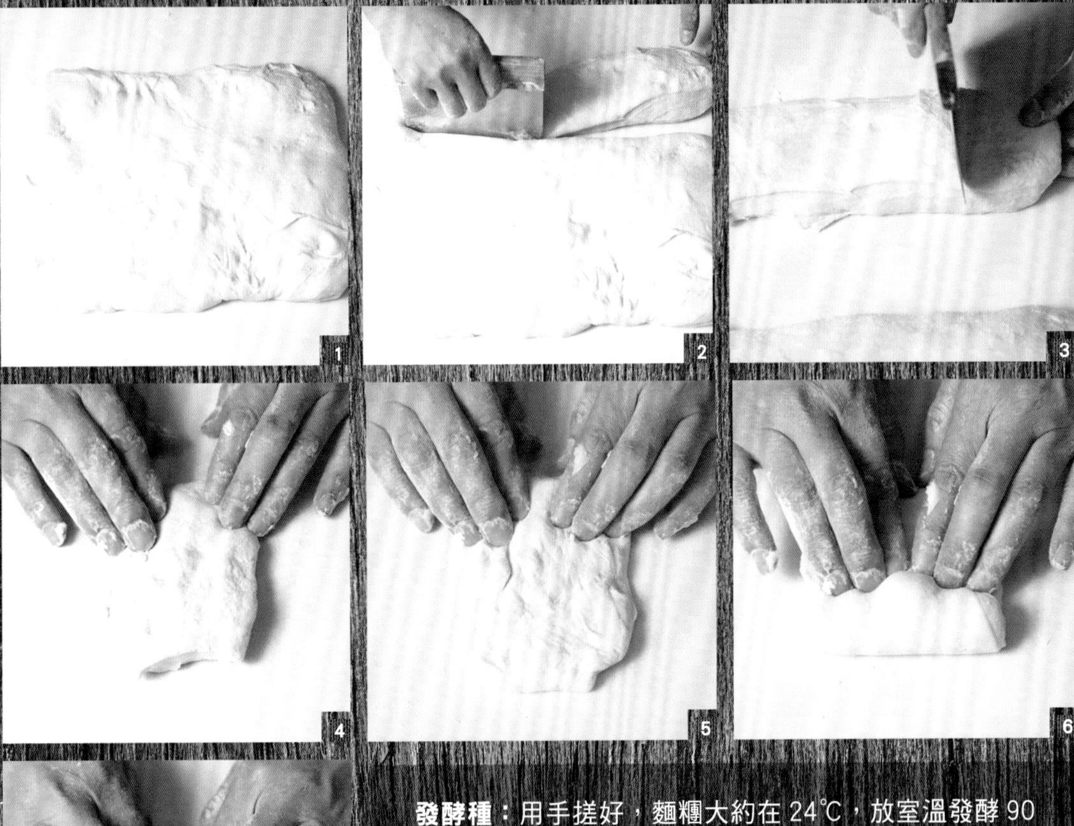

發酵種： 用手搓好，麵糰大約在 24℃，放室溫發酵 90 分鐘後放 0-5℃雪櫃 6-8 小時，或搓好放 0-5℃雪櫃冷藏 8-24 小時。

混合： 將鮮酵母溶於食譜份量的部份水中或灑在麵粉內，在大碗內把酵母水放入麵粉、餘下水、發酵種、麥芽精和鹽，以膠刮混合至不見到水份。

搓揉： 搓揉至麵糰光滑，量度麵糰溫度約 24℃。

發酵： 麵糰發酵 90 分鐘後，將麵糰拿出來放枱面上翻面後再發酵 45 分鐘（翻面做法和圖看第 72 頁「全麥包」的圖 6-11）。

排氣、分割： 第一次發酵完成，把麵糰刮出來，弄成方形，切成長條形再切細成大約 60 克一個，Handsquare 及滾圓，放膠盒內，用蓋蓋着，讓麵糰鬆弛約 15 分鐘。（圖 1 - 7）

小法包造型：桌上撒少許手粉，把 Handsquare 麵糰輕輕排出氣體，將麵糰輕拍成長方形，將 1/3 部份向下摺，倒轉，將 1/3 部份向上摺，收口摺緊成長欖形，發酵布篩粉防黏，造型後麵糰放發酵布，最後發酵約 50-60 分鐘，最後發酵完畢後，將麵糰用木棒挑到已墊不沾布的焗盤上，在麵糰上篩少許筋粉，用刴刀刴上花紋。（圖 8 - 21）

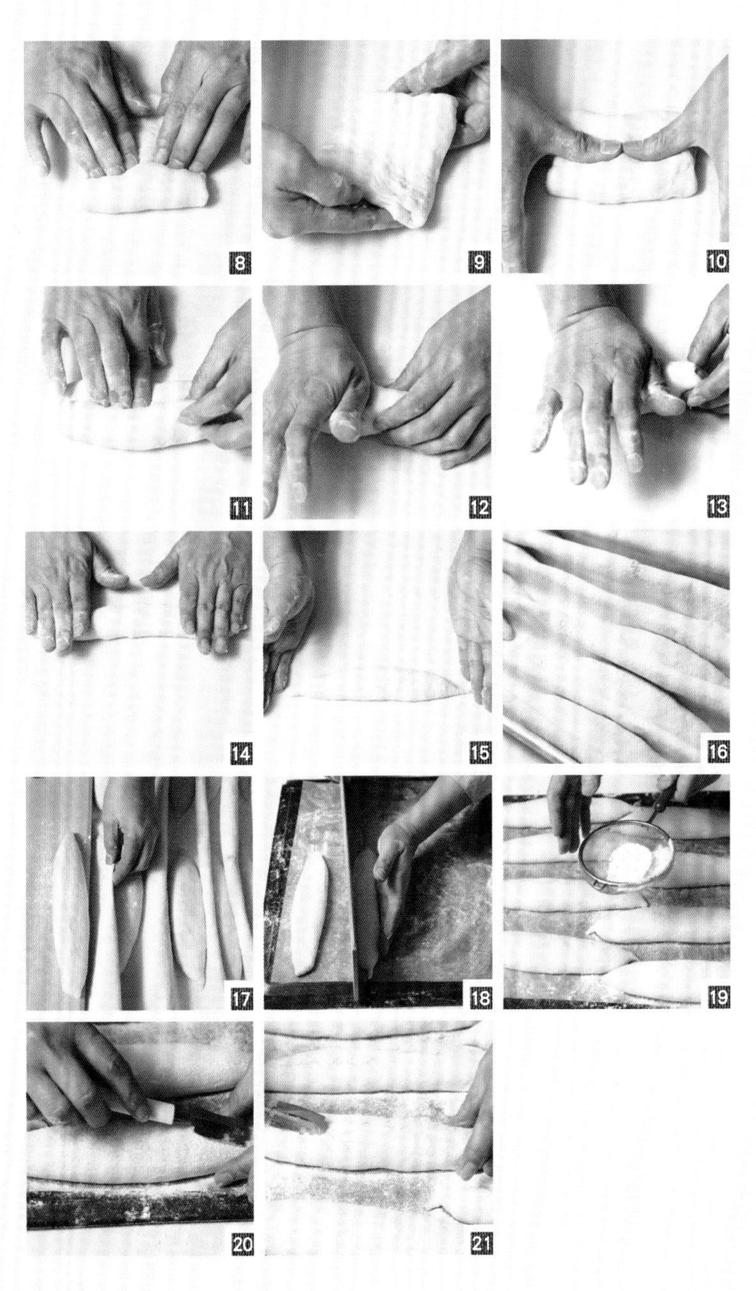

Video
小法包造型

Video
小法包刴花

麥穗包造型：把 Handsquare 麵糰輕輕排出氣體，將麵糰輕拍成長方形，將 1/3 部份向下摺，倒轉，將 1/3 部份向上摺，收口，輕搓成長條形，放不沾布或牛油紙上發酵約 50-60 分鐘，用剪刀剪成麥穗形。（圖 22 - 26）

22

23

24

25

26

雙胞胎包造型：把滾圓麵糰輕輕按扁排出氣體，幼身麵棍沾粉防黏，壓在麵糰中間，上下滾動數下，壓至麵糰中間約有 2 厘米距離，約 A4 紙厚薄，兩頭捏尖成雙胞胎形，放不沾布或牛油紙上，最後發酵 50-60 分鐘。（圖 27 - 30）

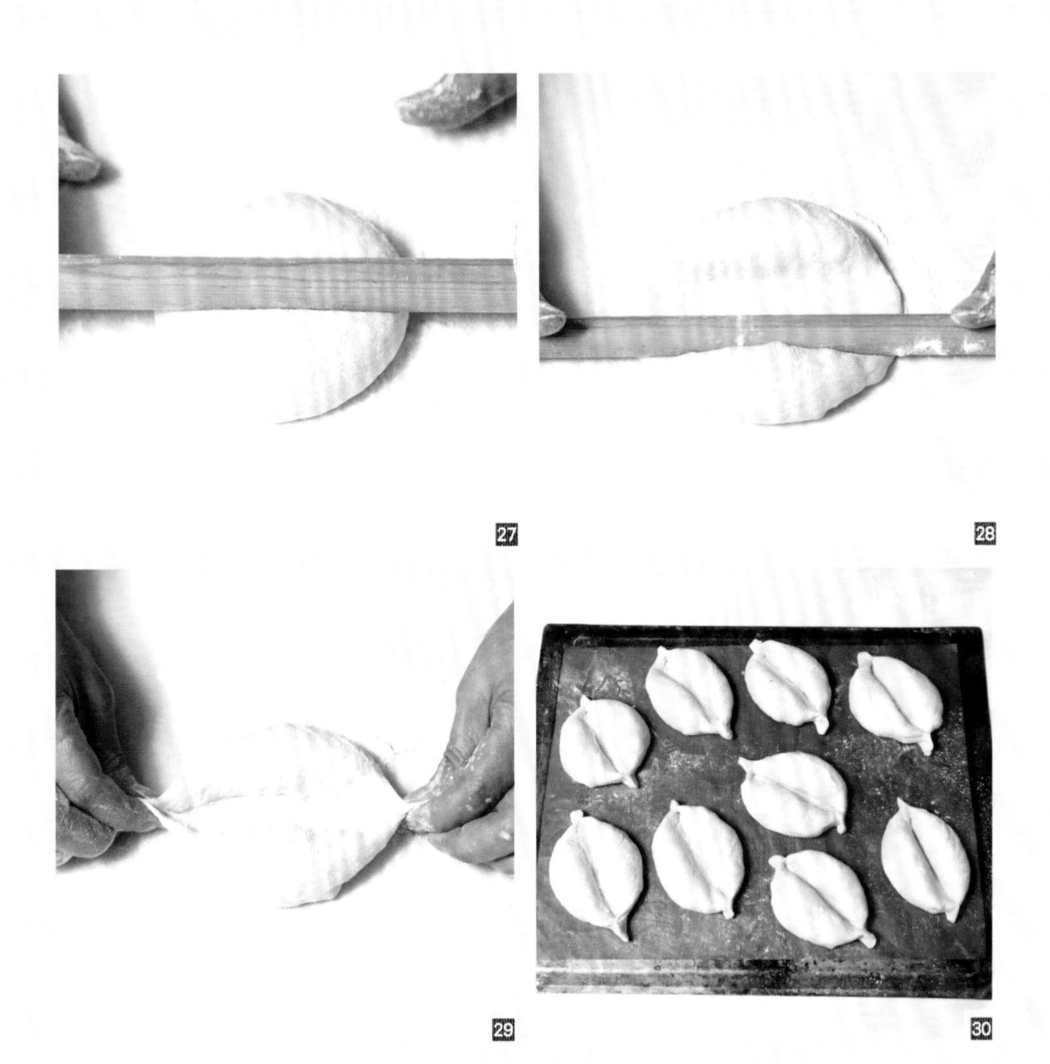

27 28 29 30

烤焗：焗爐預熱到最高溫度，連焗盤也一起預熱，預備壓縮噴水壺。打開焗爐門，立即將放有麵糰的不沾布推到燒熱的焗盤上，噴水 3 秒，關門焗約 16 分鐘，如麵糰上色後可調低焗爐溫度。

做法包心得

硬質麵包要做到外皮酥脆，是要在麵包剛入爐烤焗時在麵糰表面噴一層細如霧的熱蒸汽，如果用冷水及普通噴壺噴出來的水珠太大粒，麵糰表面的水汽過多，麵包皮就會是硬脆，會弄損口腔。

可是一般的家用焗爐並沒有打蒸汽設備，而有此設備的又體形龐大及超昂貴，為了解決這問題，可以用以下幾個方法，但要注意一下自己的焗爐門邊要有膠邊，可令水蒸汽保持才能使用，如果沒有膠邊，就要用一條半濕毛巾包着門邊進行，但效果稍遜。

其次是，硬質麵包低糖低油，要用很高溫才焗出有甘香風味，家用焗爐爐溫最高大約 250-280℃，恒溫亦沒有做得很好，打開門時溫度會下跌幾十度，關門後也很難很快升溫，為了保持高溫，要把溫度調到最高及提早預熱。

31

熱盤法

焗爐預熱到最高溫度，連焗盤也一起預熱，預熱壓縮噴蒸汽壺（是一種清潔廚房或車子的電器），打開焗爐門立即將放有麵糰的不沾布或牛油紙推到燒熱的焗盤上，在麵糰上噴蒸汽 3-5 秒。（圖 31）

熱石板法

訂購一塊焗爐石板，把石板和焗盤一起預熱到最高溫度，預熱壓縮噴水壺，打開焗爐門，立即將放有麵糰的不沾布或牛油紙推到石板上，在麵糰上噴蒸汽 3-5 秒。

32

33

鑄鐵鍋烤焗法（圖 32 - 33）

將鑄鐵鍋連蓋放焗爐一起預熱，將放有麵糰的不沾布或牛油紙小心放到燒熱的鑄鐵鍋內，蓋上鍋蓋，烤焗約一半時間，打開蓋讓麵糰上色至熟透。因為入麵糰後，鍋內能把麵糰釋放的水汽鎖住，遇熱後形成一層水蒸汽在麵糰上，麵包爐爆很好。

所謂爐爆是麵糰入爐後膨脹後的割口裂開的狀況。

熱石灒水法

把鵝卵石用一小盤盛着放爐最下層，一起預熱到最高溫度，用熱盤法或熱石板法，放入麵糰後用一杯熱水灒下熱石製造水蒸汽，注意水不能過多，水份過多底火就會變弱。底火是很重要的一環，底火不足會影響爐爆。

熱濕毛巾法

用熱盤法或熱石板法，在爐底也放一小焗盤，預熱到最高溫度放入麵糰，以一條濕了熱水的毛巾放爐底小焗盤讓其釋放蒸汽。蒸汽足夠就把毛巾移開。

French Bread

Fermented dough method: Baton, Pain D'Epi and Twins

*Fermented dough (pâte fermentée)	Baker's Percentage (%)	Weight (g)
French flour Type 55	100%	100 g
sea salt	2%	2 g
malt extract	0.5%	0.5 g
fresh yeast	0.5%	0.5 g
water	65%	65 g

Main dough		
French flour Type 55	100%	300 g
malt extract	0.5%	1.5 g
sea salt	2%	6 g
fresh yeast	2%	6 g
water	66%	198 g
old dough	15%	45 g
	185.5%	**556.5 g**

Fermented dough

Knead the ingredients until the dough reaches 24°C. Leave it at room temperature for 90 minutes. Transfer in a fridge at 0 to 5°C and keep it there for 6 to 8 hours. Alternatively, knead the ingredients until well incorporated and refrigerate at 0 to 5°C for 8 to 24 hours.

Mixing

Dissolve the fresh yeast in part of the water first. Or sprinkle fresh yeast in the bread flour. Transfer the yeast mixture into a mixing bowl. Add bread flour, the remaining water, fermented dough, malt extract and salt. Stir with a rubber scraper until no dry patch is visible.

Kneading

Knead the dough until smooth and until it reaches 24°C.

First proofing

Leave the dough to proof at room temperature for 90 minutes. Punching down the dough on a countertop. Let it proof for 45 more minutes. (Refer to photo 5-11 in the recipe "Wholemeal loaf".)

Punch down / dividing / final proofing

After first proofing, scrape the dough out and shape it into a square. Cut into long strips and then into pieces about 60 g each. Handsquare them each or roll them round depending on the shape you intend to make. Put them into a plastic box and cover the lid. Let them rest for 15 mintues. (pics. 1 - 7)

Shaping

- **Baton:** Sprinkle some bread flour on the countertop. Gently press each handsquared dough piece to release air bubbles. Gently press the dough into a rectangle, with the short side towards you. Fold the top third along the length downward. Then flip it upside down and fold the bottom third upward along the length. Pinch to seal the seams. Shape into an oval shape with pointy ends. Transfer the dough pieces onto a piece of floured proofing cloth. Leave them to rise once more for 50 to 60 minutes. Transfer the dough pieces using a wooden bread peel onto a baking tray lined with silicone baking mat. Sieve some bread flour over the batons. Make slashes with a bread lame. (pics. 8 - 21)
- **Pain D'Epi:** Gently press each handsquared dough piece to release air bubbles. Gently press the dough into a rectangle with the short side towards you. Fold the top third along the length downward. Then flip it upside down and fold the bottom third upward along the

length. Pinch to seal the seams. Roll the dough on the counter into a log. Transfer onto a silicone baking mat or parchment paper. Leave it to proof for 50 to 60 minutes. Cut the dough into wheat stalk shape with scissors. (pics. 22 - 26)

- **Twins:** Gently press the rolled round dough piece to release air bubbles. Flour a thin rolling pin. Put it at the centre of the dough and roll back and forth a few times until the indentation is 2 cm wide and paper thin. Pinch the both ends of the indentation into pointy protrusions. Transfer onto a silicone baking mat or parchment paper. Leave it to proof for 50 to 60 minutes. (pics. 27 - 30)

Baking

Preheat an oven together with the baking tray to the highest temperature setting available. Have a compression steam bottle ready. Open the oven door and quickly transfer the dough pieces and the silicone baking mat onto the hot baking tray. Spray water mist for 3 seconds. Close the oven door and bake for 16 minutes. Once the bread starts to brown, turn down the oven.

Tips on making crusty French bread

- Crusty bread, as its name suggests, should have thick and crispy crust on the outside. The key is to create an atmosphere of fine water mist on the surface of the dough when it goes into the oven. If you use cold water and regular spray bottle, the droplets will be too large and too much water accumulates on the dough. The crust will turn out too hard and stiff that way, cutting the roof of your mouth.

- However, most home-scale ovens don't come with steaming function. Those that do are always bulky and cost a fortune. Here are a few solutions to create steam without breaking the bank. But you should check if your oven has rubber seals along the door first. The rubber seals are essential to keep the steam in. Without rubber seals, you may still bake with a half-damp towel wrapping along the gap between the door and the frame. Yet, it never works as well as the rubber seals. The second issue is the temperature. Crusty bread must be low in sugar and grease, and bake at a very high temperature for the right texture to develop. Most home-scale ovens reach 250°C or 280°C at most. They usually suck at retaining heat – once you open the door and the temperature drops tens of degrees instantly. After you shut the door, the temperature isn't going to rise very quickly either. To retain the strong heat, you should preheat the oven in advance to its highest temperature possible.

Heated baking tray

Preheat the oven and the baking tray to the highest temperature possible. Have a compression steam bottle preheated – it is an electrical appliance for cleaning kitchen or cars. Open the oven door and quickly transfer the dough on silicone baking mat or parchment paper onto the baking tray. Spray steam on the dough for 3 to 5 seconds.
(pics. 31)

Heated baking stone

Purchase a baking stone. Preheat the oven and the baking stone to the highest temperature possible. Preheat a compression steam bottle. Open the oven door and quickly transfer the dough on silicone baking mat or parchment paper onto the hot baking stone. Spray steam on the dough for 3 to 5 seconds.

Cast iron pot

Preheat the oven and the cast iron pot with lid on. Open the oven door and carefully transfer the dough on silicone baking mat or parchment paper into the cast iron pot. Cover the lid. When the bread has been baked half-way, over the lid and let the bread brown. Cast iron pot works as the lid seals in the steam after the dough is put in. A layer of steam is kept on the dough so that it is known for excellent oven spring and opening up slashes nicely. (pics. 32 - 33)

Sprinkling water on heated pebbles

Put clean pebbles on a small oven-safe tray. Put it on the lowest rack in the oven. Preheat the oven together with the baking tray or baking stone. After you put in the dough, pour a cup of boiling water onto the pebbles to create steam. Make sure you don't pour in too much water though. Otherwise, the water would weaken the bottom heat. The bottom heat is essential to baking bread. Insufficient bottom heat would retard oven spring.

Hot wet towel

Follow the heated baking tray or heated baking stone method above. Then put a small baking tray on the bottom of the oven. Preheat the oven to the highest temperature possible and put in the dough. Immediately put a hot wet towel on the bottom tray to create steam. When there is enough steam, remove the towel.

粟米沙律包

直接法：翻面
學習製作不同餡料

麵糰材料	百分比（%）	重量（克）
高筋麵粉	90%	270 克
全麥麵粉	10%	30 克
砂糖	10%	30 克
海鹽	2%	6 克
脫脂奶粉	3%	9 克
蛋	10%	30 克
蛋黃	8%	24 克
鮮酵母	3%	9 克
水	55%	165 克
無鹽牛油	12%	36 克
	203%	**609 克**

飾面

車厘茄

Mozzarella 芝士

麵糰做法：參考第 38 頁「圓形小餐包」的混合、搓揉、投油、測試筋膜和發酵的步驟和圖 1-25。

麵糰發酵 60 分鐘後，將麵糰拿出來放枱面上翻面，再發酵 30 分鐘。（翻面做法看第 74 頁「全麥包」的圖 5-11）

排氣、分割：排氣、分割成 60 克麵糰，滾圓（參考第 41 頁「圓形小餐包」內的排氣、分割、滾圓的步驟和圖 30-38），放雪櫃休息 15-20 分鐘至麵糰大約 15℃。

造型：參考第 69 頁「軟小餐包 - 編織包 5 號花造型」的步驟和圖 16-21，放紙兜內發酵。

最後發酵：大約發酵 20-25 分鐘，將餡料用雪糕杓舀到麵糰上，用少許力壓下餡料，在餡料上放半粒車厘茄，再放上 Mozzarella 芝士。（圖 1 - 3）

焗焗：放入已預熱約 200℃焗爐，焗約 13-14 分鐘。

4

5

6

粟米沙律製作方法

材料	
罐頭粟米粒	250 克
洋葱粒	20 克
火腿粒	30 克
熟蛋	2 個
沙律醬	60 克
海鹽、胡椒	適量
雜芝士碎（薄餅芝士或用車打芝士）	50 克

做法：洋葱粒用油爆香，熟蛋切碎，與罐頭粟米粒、火腿粒、雜芝士碎、沙律醬、海鹽和胡椒混合。（圖 4 - 6）

Sweet Corn Salad Buns

Direct method: Punching down.

Learning about making dfferent filling.

Main dough	Baker's Percentage (%)	Weight (g)
bread flour	90%	270 g
wholemeal flour	10%	30 g
sugar	10%	30 g
sea salt	2%	6 g
skimmed milk powder	3%	9 g
egg	10%	30 g
egg yolk	8%	24 g
fresh yeast	3%	9 g
water	55%	165 g
unsalted butter	12%	36 g
	203%	**609 g**

Garnish
cherry tomatoes
grated mozzarella cheese

Making the dough

Follow the steps in the recipe "Plain Round Rolls" for mixing, kneading, adding butter, windowpane test and first proofing.

After the dough has proofed for 60 minutes, transfer the dough onto a countertop and punching down. Leave it to proof for 30 more minutes. (Refer to photo 5-11 in the recipe "Wholemeal loaf".)

Punch down & dividing

Press the dough to release air bubbles. Divide into dough pieces about 60 g each. Roll each round (follow the steps of the recipe "Plain Round Rolls" for punch down, dividing and rolling round). Leave them to rest in a fridge for 15 to 20 minutes until the dough temperature comes down to 15°C.

Shaping

Follow the steps in the recipe "Braided bread #5". Put the dough pieces into paper baking cups for final proofing.

Final proofing

Leave them for 20 to 25 minutes. Scoop the sweet corn topping on each bun. Gently press the topping. Arrange 1/2 a cherry tomato over. Top with grated mozzarella cheese. (pics. 1 - 3)

Baking

Bake in a preheated oven at 200°C for 13 to 14 minutes.

Sweet corn topping

Ingredients

canned sweet corn kernels	250 g
diced onion	20 g
diced cooked ham	30 g
hard-boiled eggs	2
creamy salad dressing	60 g
sea salt	
ground white pepper	
assorted grated cheeses (pizza topping cheeses or cheddar cheese)	50 g

Method: Stir-fry onion in a little oil until fragrant. Set aside. Dice the eggs. Put all ingredients into a mixing bowl and mix well. (pics. 4 - 6)

初嘗麵包香 Breadmaking for the Beginners

作者 Author
獨角仙 Kin Chan

策劃/編輯 Project Editor
譚麗琴 Catherine Tam

攝影 Photographer
Yan Kin-wai, Chiu Yuk-shing

美術設計 Design
鍾啟善 Nora Chung

出版者 Publisher
Forms Kitchen
香港鰂魚涌英皇道1065號 東達中心1305室
Room 1305, Eastern Centre, 1065 King's Road, Quarry Bay, Hong Kong.
電話 Tel: 2564 7511
傳真 Fax: 2565 5539
電郵 Email: info@wanlibk.com
網址 Web Site: http://www.wanlibk.com
http://www.facebook.com/wanlibk

發行者 Distributor
香港聯合書刊物流有限公司
SUP Publishing Logistics (HK) Ltd.
香港新界大埔汀麗路36號 中華商務印刷大廈3字樓
3/F., C&C Building, 36 Ting Lai Road, Tai Po, N.T., Hong Kong
電話 Tel: 2150 2100
傳真 Fax: 2407 3062
電郵 Email: info@suplogistics.com.hk

承印者 Printer
中華商務彩色印刷有限公司
C & C Offset Printing Co., Ltd.

出版日期 Publishing Date
二零一九年四月第一次印刷
First print in April 2019

Published in Hong Kong by Forms Kitchen,
a division of Wan Li Book Company Limited.
ISBN 978-962-14-6923-6